Blood On My Hands

a Haematological Odyssey

by

Dr Bryon E Roberts

MD FRCP Edin FRCPath

formerly

Consultant Haematologist

The General Infirmary at Leeds

Copyright © 2014 Dr Bryon E Roberts

All rights reserved, including the right to reproduce this book, or portions thereof in any form. No part of this text may be reproduced, transmitted, downloaded, decompiled, reverse engineered, or stored, in any form or introduced into any information storage and retrieval system, in any form or by any means, whether electronic or mechanical without the express written permission of the author.

ISBN: 978-1-291-71767-9

Front Cover: Bone marrow in hairy cell leukaemia.

PublishNation, London
www.publishnation.co.uk

To Audrey and David who for several years did not know where my career was going; this is hardly surprising because neither did I.

CONTENTS

PREFACE

When Graeme Smith, a Senior Consultant Haematologist, asked me to write a history of haematology in Leeds I anticipated that it would be objective and written in the third person. It soon became obvious that the only source of the information in my early years was my memory and so this book is written in the first person because all the actions and decisions in haematology for a long time were mine and there is no paper record.

When Harold MacMillan was Prime Minister he was asked by a journalist what was likely to blow the government off course and his reply was: "events, dear boy, events".

This account of the development of haematology in Leeds is the story of a series of events and the reaction to them. Rarely does this story have a clear progressive theme.

During my school days when contemplating a career in medicine came the first major event; the creating of the National Health Service in 1948. I entered Medical School in 1951 and I describe life in a provincial major teaching hospital in the years following, both as a medical student and then doctor. I decided upon a post graduate career in haematology a specialty which then did not exist except in a few teaching hospital departments of medicine. My post graduate training would be entirely organised and financed by myself. After a few years in an unhappy University Department of Pathology getting nowhere, I was lucky to be offered a registrar post at the Royal Post Graduate Medicine School at the Hammersmith Hospital in the Department of Haematology which was world renowned. This was my road to Damascus where I received a Pauline conversion. I found out what it was like to practice high quality academic medicine and I promised myself that I would never deviate from these standards.

The next major event was the creation of the Royal College of Pathologists who divided Pathology into the major specialties and thus the Specialty of Haematology was established. When I obtained the examination for Membership of the Royal College I knew my future career was safe.

I became a National Health Service Consultant in Leeds in 1970 when life became a maelstrom of events local, national, and international, all of which would affect the future of haematology.

Locally in Leeds there was intensive rivalry between the Infirmary and its sister hospital St James's across town and both hospitals hated Bradford. This was much to the detriment of medicine in Yorkshire and over much of my career in Leeds. The relationship between the Hospital and University was a troubled one.

At a national level the country seemed to lurch between boom and bust which has continued to this day. One could plan a new hospital one year, or a new consultant post, only to find them cancelled the next. Successive governments would all make changes to the National Health Service, some beneficial others not, but all changes would have repercussions for the development of haematology. Overall there were threats and there were opportunities and one had to take the opportunities when the arose.

Throughout the time covered by this book from 1951 to 1998 there were major epoch making developments in the medical and biological sciences and all, it seemed, would change the nature of haematology practice.

In the late 1930's and early 1940's the pathogenesis of Pernicious Anaemia was discovered and the missing nutritional factor Vitamin B12 synthesised. This was followed by the discovery of Folic Acid which was another nutritional factor metabolically linked to Vitamin B12. Pernicious Anaemia a hitherto fatal disease could now cured by regular Vitamin B12 injections. This provoked a vast amount of research in haematology into the nutritional anaemias in which I was involved in my early years.

In 1953 Crick and Watson in the Laboratory for Molecular Biology at Cambridge described the structure of DNA using x-ray crystallography. They had the help of Franklin and Wilkins from King's London who provided crucial radiographs. This discovery was as important as Darwin's theory of evolution and in fact confirmed it. Its impact in haematology was not felt until the 1980's when recombinant DNA technology and gene sequencing were introduced.

In the 1960's tissue typing was developed to facilitate organ transplantation. Bone marrow transplantation would follow later when developments in chemotherapy led to remission induction which was essential before bone marrow transplantation could take place.

In the late 1960's and early 1970's new chemotherapeutic agents were introduced. When we learned to use them in combination and at intervals the treatment of leukaemia was transformed as was indeed the practice of haematology.

In 1973 Cesar Millstein developed monoclonal antibodies which were specific to a particular biological substance. These are expensive but have transformed the therapy of disease, such as rheumatoid arthritis. They have made a major contribution to the diagnosis and classification of haematological malignancies and form the basis of new chemotherapeutic agents.

The story of haematology is therefore that of some of the most exciting scientific and therapeutic developments in medicine all of which have transformed haematology practice and made Haematology a very exciting subject to practice.

In Chapters 4 and 6 there are detailed technical descriptions of some of the diagnostic tests used at the time. They do disturb the narrative flow but to me have historical importance. The non haematological reader should pass them by.

Bryon Roberts
Leeds 2013

ACKNOWLEDGEMENTS

The discipline of Haematology has had a prolonged and difficult gestation in Leeds. It has, now that it is situated in the Bexley Wing, a centre for Oncology, reached maturity, and acquired an international reputation.

Dr Graeme Smith, a Consultant Haematologist in the Department of Haematology, thought it a good time to write an account of the development of Haematology in Leeds and asked me to do this as I was associated with Haematology in Leeds from the outset. I discussed this further with Andrew Bannister, Head of Media Relations, The Leeds Teaching Hospitals NHS Trust. It was he who suggested that I broadened the scope of the book, which I have done. In particular I have used my memory to describe the early days of the NHS in Leeds.

The NHS is subject to cataclysmic changes in management from time to time which means records are difficult to find, so much of the book is written from both my memory and that of others I have chosen to consult.

In particular Bill Mathie, recently Secretary to the Medical School, has been a very useful source of information, as has the annual periodical Medicine Matters, an alumni magazine of the Medical School. Published by Leeds University and edited until recently by Bill there are many records of events and personalities of historical significance. Bill has read the first draft of the book and made useful comments.

I am also greatly indebted to Dr Malcolm Parsons, formerly Consultant Neurologist to the General Infirmary at Leeds, and an almost exact contemporary of mine. He has written two excellent books on the history of medicine in Leeds and Yorkshire. He read the first draft of the book and made several very helpful and constructive comments.

I must thank Dr Roger Owen for taking the photographs of the antique items of haematological equipment and for converting several scrappy and elderly photographs in a reasonable digital format.

Philip Day and Geoffrey Tate, former Chief Technicians in the Department of Haematology, have been very helpful in enabling me to recall various incidents and people.

Finally I must thank Mrs Louise Law who has converted my dictation and scribble into a reasonable format. Her speed and accuracy have been beyond reproach and I am greatly indebted to her.

The publications I have consulted are as follows:-

Dacie JV and Lewis SM - *Practical Haematology* - Churchill & Livingstone, London, 1984

Porter Roy - *The Greatest Benefit to Mankind* - Harper Collins, London, 1997

Parsons Malcolm - *Yorkshire and the History of Medicine* - Sessions of York, 2002

Leukaemia Research Fund Centre for Clinical Epidemiology University of Leeds: *Leukaemia and Lymphoma - an atlas of distribution within areas of England and Wales 1984-88* - Leukaemia Research fund 1990.

Marr Andrew - *A History of Modern Britain* - MacMillan, London, 2007

Ward AJ and Ashton T - *Cookridge Hospital 1867-1972* - University of Leeds, 1997

The web-site of the Leeds Institute of Medical Education has several important biographies of Leeds medical staff of distinction which have been consulted.

Bryon Roberts
2013

"All men dream, but not equally. Those who dream by night in the dusty recesses of their mind, wake in the day to find it was vanity; but the dreamers of the day are dangerous men. For they may act on their dreams with open eyes to make them possible."

T E Lawrence

CHAPTER 1

IN THE BEGINNING ...

"A prophet is not without honour but in his own land."
Mark 6:4

Charles Turner Thackrah was born in Shadwell in 1795. He was subsequently apprenticed at the General Infirmary at Leeds and then went to Guy's Hospital where he was a contemporary of John Keats, the poet. He was taught by the renowned Astley Cooper and was awarded the Astley Cooper prize (this experience would benefit him greatly when he opened his anatomy school in Leeds).

Thackrah returned to Leeds and, as a man of culture, was invited to give the inaugural lecture of the Leeds Philosophical and Literary Society. He was not, however, elected to a post at the Infirmary by subscribers but instead, much to his dismay, had to take a post of town surgeon with a much lower salary, but this was in fact to provide invaluable experience for his future observations on the association of occupation and disease.

He was, however, a distinguished scientific investigator and finally decided to open a School of Anatomy of his own in South Parade in 1826. This school was popular and it is likely that it was supplied with bodies by resurrectionists, or body snatchers. Inevitably this provoked controversy and conflict with the Infirmary, particularly the surgeon, Samuel Smith, which culminated in an acrimonious series of letters in the *Leeds Mercury*. As part of this conflict it was revealed that Thackrah had an illegitimate son following his seduction of a young girl who was his patient.

However, new schools of medicine were being established in the provincial cities of England and on the 6th June 1831 six surgeons and physicians resolved on the establishment of a medical school in Leeds. Thackrah was sent an invitation to cooperate, which he accepted, and he closed his School of Anatomy. Leeds Medical School was opened some four months later and Thackrah gave lectures to the medical students on anatomy, physiology, pathology, and surgery for the next two years.

Due to his experience as town surgeon, Thackrah began to express grave concerns about child labour and injuries arising from industrial working conditions in and around Leeds, and Thackrah's reputation as one of the founders of English provincial medical education is only surpassed by his reputation as the father of occupational medicine in the English speaking world. Thackrah suffered with ill health, in particular he had severe ulcerative colitis but, nevertheless, brought forth a first edition of his work on industrial diseases in 1831. His book was a triumph and was reprinted in America almost immediately. A large and definitive edition followed in 1832 entitled *"The Effects of Arts, Trades and Professions and of civic states and habits of living on Health and Longevity with suggestions for removal of many of the agents which produce disease and shorten the duration of life"*. The book's strength was its breadth of coverage of over a hundred trades in Leeds at the time. It described among other things the postural deformities in child mill workers, dust diseases in miners, and among those exposed to harmful substances were corn millers, maltsers, coffee roasters, snuff makers, rag pickers, paper makers, and feather dressers; tailors were so subject to anal fistulas that they set up their own fistula clubs. Thackrah went on to claim that not ten percent of the inhabitants of large towns enjoy full health. He made important recommendations for the prevention and considered "thoughtlessness or apathy is the only obstacle to success in the removal of injurious agents".

Occupational Medicine is a discipline which was established as a result. His work contributed to the passing of the Factory Act in 1833 which prohibited the employment of children under nine years old in the textile mills. Thackrah died of tuberculosis in 1833, at the age of thirty-eight. The first Chief Medical Officer in 1855, Sir John Simon, considered Thackrah's contribution to preventative medicine as comparable to the work of Jenner on smallpox.

As a medical student and throughout my professional career, I knew of Thackrah as the father of occupational medicine, but he was never, for example, a hero figure like Moynihan.

On my retirement I was asked to give a lecture to the Medical and Dental History Society on the history of haematology. I agreed but then realised that my knowledge of the history of haematology was somewhat patchy, but I was fortunate to find the excellent book on the topic by Maxwell Wintrobe, an American haematologist of great international distinction, entitled *"Blood, Pure and Eloquent: A Story of Discovery, of People, and of Ideas"*. Maxwell Wintrobe was influential enough to be able to attract some of the most distinguished workers in their field to contribute chapters. I had no option but to read the book cover-to-cover.

During my reading of the chapters on blood coagulation, mainly by American authors, I kept coming across a reference to C. T. Thackrah. I thought at first that there must be an American with the same name; surely it could not be our very own Leeds', C. T. Thackrah - but it was. There is a list of references cited at the back of the book according to author. Thackrah was well in the top half of the list of authors outscoring many haematologists of repute. I was considerably disturbed to find that having been a student, and then having practised and taught haematology in Leeds for forty years, I had not known this and had never been able to inform the students that Thackrah, one of our founding fathers, was a pioneer in the study of haemostasis for which he had a world wide reputation.

Virtually all patients attending medical clinics from the time of the ancient Greeks to the late nineteenth century were treated by bleeding. After the application of a tourniquet, the vein was nicked by a scalpel and a jet of blood emerged to be collected in a pot.

3

What Thackrah did was to collect blood sequentially into Galley pots and measure the clotting time. The blood was allowed to retract and the clot, or crassamentum as Thackrah called it, was weighed. Thus Thackrah was able to establish a normal range for what we now know as fibrinogen. That normal range is the one in use today.

Among other observations that he made were: that different surfaces affect the speed of clotting, that blood clotted slowly in a legated vein, that tissue accelerated blood clotting, and that there was delayed blood clotting in liver failure. Much of this can be found in a monograph written in 1819 by Thackrah entitled *"An Enquiry into the Nature and Properties of the Blood"*. This monograph was published shortly afterwards in the United States. A copy of this monograph is situated in the Brotherton Library and I visited the library to read it. There are many case descriptions and I was appalled to read many of them.

At the time physicians had no concept of anaemia and therefore patients coming into the clinic with what appeared to be a haemorrhagic disease associated with perhaps acute leukaemia were treated by venesection and probably bled to death. I was so shaken by what I read that I was unable to finish the monograph.

Thus, Thackrah established himself as a pioneer in the field of blood coagulation and, in many ways, anticipated the work on the fibrogen-fibrin reaction that Leeds has been so closely associated with in recent years.

Thackrah was not only the father of occupational medicine but also one of the first doctors to perform investigations in coagulation in the United Kingdom.

CHAPTER 2

THE TWILIGHT OF THE GODS

I came to Leeds Medical School in 1951 three years after the NHS was instituted in 1948.

The industrial cities of Britain were grim and gloomy places. There was smoke everywhere from the factories, coal fires, and tobacco. The buildings were black and encrusted by soot and, as I was to observe later, so were the lungs of the inhabitants when seen at post mortem. The unremitting gloom was alleviated as one walked to the Infirmary by the gleam and glitter of the chromium radiators, of the Rolls Royces and Bentleys of the Consultant staff. Aneurin Bevan declared that he would ensure the success of the NHS by "stuffing the mouths of consultants with gold". By the look of the cars many of the owners were not short of the odd ingot or two already.

Clothes and food rationing had only just finished. Everyone was dressed in drab clothing; the men, if not wearing a suit, would wear grey flannels, black shoes, and a sports coat covered by a Mac. The wearing of a hat or cap was universal. There was no domestic central heating and men frequently wore long johns and their stockings were held up by suspenders. The shirt had a detached collar which was held to the shirt by collar studs. Suits were often worn at the weekend. Leeds was a centre of the manufacture of ready made clothing. In particular, Montague Burton had a very large factory closely situated by St James's Hospital on the other side of the cemetery. There were other manufacturers, such as, The Fifty Shillings Tailors. One could buy a complete outfit at Montague

5

Burton's, in other words "the full Monty". There were two middleclass department stores in Leeds, Schofields which catered mainly for the middle class, and Marshall & Snelgrove's an upper crust department store, part of a national chain.

Marshall & Snelgrove's was frequented by many consultants. When Digby Chamberlain, the Leeds Surgeon, visited Marshall & Snelgrove's with his wife it was an event to be seen. He would draw up in Bond Street, which was then not pedestrianised, the uniformed chauffeur would leave the car and open both rear doors to allow Mr and Mrs Chamberlain to alight thus blocking all the traffic. Digby, and his wife, would then walk into Marshall & Snelgrove's with a regal air and the normal traffic flow in the street would be allowed to restart.

Many consultants went to have their hair cut in the basement of Marshall & Snelgrove's. The chairs were arranged in a circle with the customer looking outwards. The barbers would cut the hair from the back and at the front attractive young women would perform a manicure. It is said that some of these young women would produce other services, out of hours, for a suitable fee.

As medical students we did not buy our scarves or blazers from the same shop as the rest of the University students. We went to Marshall & Snelgrove's and were suitably attired in a blazer which was maroon with stripes of light blue and pink, a similarly patterned cravat for when we had an open necked shirt, a medical school tie, and scarf.

The Medical Faculty with the Law Faculty were the largest in the University. In general, in the University, the medical students were not liked and, in some ways, the medical students considered themselves superior to the others since the Medical Faculty existed before the University was founded in the early years of the twentieth century.

The first six months of our study was in the University bringing all the students in our year of varying backgrounds up to speed. We also studied Medical Statistics.

The next two years were devoted to the study of Anatomy, Physiology, and Biochemistry. We would have lectures at 9.00 am and 12 noon, and often at 4.00 pm. The time in between was spent in the dissecting room and the laboratory. In anatomy we proceeded to dissect a cadaver. One pair of students would dissect one side and another pair the other. At school I had dissected a cockroach, a worm, a frog, a dogfish, and a rabbit, and now I was dissecting the top of the evolutionary tree. All students had to buy a microscope and half a skeleton, which I kept under my bed.

In physiology we would perform experiments on frogs. We would pick up a live frog by its back legs and hit its head on the edge of a bench to render it unconscious. We then did experiments on the function of muscles and the nerves by connecting the leg to a pointer which marked a rotating smoked drum. Thus we were able to provide tracings on the drum of the functions of the muscles and nerves when subjected to various stimuli. On another occasion a live dog was anaesthetised, the chest opened, and the physiology of the heart demonstrated before the dog was sacrificed.

One January morning in the physiology practical period we performed an experiment in haematology. Despite our very cold fingers we pricked them and sucked a sample of blood up a pipette which we blew into a diluent. We connected a tube to a gas tap and blew coal gas through the dilute haemoglobin, converting it into carbonmonoxyhaemoglobin. We compared the colour of this to a standard and read off the haemoglobin in percent of normal. We then ran up a hill to the University Union and back and repeated the process so that we could find out what changes of haemoglobin there was in response to exercise. The results were then analysed by a statistical method. This was my first contact with my future speciality.

After the rigor of the second MB we looked forward to three week's holiday but found out we had to return in one week to begin

7

clinical studies. Long university holidays were gone for ever. We were then divided into groups to be taught clinical methods. My tutor was Jim Fountain. Jim was Tutor in the University Department of Medicine and had a major interest in haematology. At Edinburgh University he had obtained an MD with gold medal for studying the action of 6 mercaptopurine in leukaemic mice. Our paths would cross again. After this short course we were then divided into groups and allocated to a clinical team known as a "firm". We thus had our first contact with the National Health Service.

The NHS was under duress. The Country had needed a massive loan from the United States after the war (which was only paid off during the time that Tony Blair was Prime Minister) and then had to provide troops for NATO to counter the threat of Russia. Shortly after this the Korean War started and we, as now, found that we were fighting a war in Korea supporting the United States against China. The Government therefore had no funds whatsoever to support the newly created NHS and, in particular, there were no additional consultant posts. Life was extremely difficult for consultant trainees and many had to wait a considerable time for a consultant job, if they were lucky. Many regrettably dropped out and went into General Practice.

We therefore entered a National Health Service which had changed very little since it was instituted.

In surgery there were three firms, Digby Chamberlain and Michael Oldfield, George Armitage and Henry Shucksmith, and Johnny Latchmore and P J Moir, all protégées of Moynihan.

I was allocated to Mr Chamberlain and Mr Oldfield's firm. We had to attend the wards every day and perform various tasks, such as phlebotomy. We would attend Surgical Out Patients and the operating theatre to watch consultant surgeons at work. We were taught to scrub up and, occasionally, assisted in an operation.

When the House Surgeon was on holiday students were allowed to do locums and I spent two weeks as a Locum House Surgeon. On the wards we were under the supervision and surveillance of the Ward Sister. Should we transgress in any way, in particular in the handling of patients, then we would get a dressing down never to be forgotten, but by the time we qualified and became House Surgeons we were already quite experienced on the ward and in the theatre.

The medical firm I was allocated to was the Professorial Medical Unit under Professor Sir Ronald Tunbridge. There were several senior lecturers who also performed teaching. We attended the Ward Rounds and Out Patients. There was also one major open teaching each week in the Littlewood Hall where we had to present cases and get torn to shreds if our performance warranted it. We also had to spend time as Resident Medical Dressers, and see all the acute admissions with the House Physician. Professor Tunbridge would do ward rounds out of hours, particularly on a Saturday evening. It was said that the reason for this was that he was escaping his mother-in-law. One dark, rainy evening we went with Professor Tunbridge to see a patient being nursed outside on the veranda close to what was Ward 6, exposed to all the elements. Over her bed was a tent-like structure covered in black tarpaulin. She had pulmonary tuberculosis and fresh air was considered paramount for her recovery, as well as the streptomycin therapy she was having, a treatment which was introduced after 1947. In fact, as students we were taken to see an orthopaedic annex at Boston Spa where children and adolescents were being nursed in the fresh air for bone tuberculosis.

Among other jobs we had to do were routine venesections on the wards and side room tests. The venesections were performed with glass syringes with steel needles which were often blunt and had barbs. When we completed a venesection we would rinse the syringe out in the sink at the end of the ward and boil it up to sterilise it. We would then wait until the syringes were cool enough to handle and then perform a further venesection with a very wet and hot syringe. Laboratory results, particularly potassium levels, would have been hopelessly inaccurate. We performed many side room tests in the laboratories outside Ward 6 and Ward 2. We would centrifuge urines and look at the deposit with a primitive monocular microscope and test urine for sugar and protein (there were no paper strip tests

9

then). The laboratories were supervised by Mr Frank Leak who had performed this task for many years and for this awarded the BEM.

Medical and surgical firms lasted six months and then we would rotate round many other specialty firms for shorter periods of time. I remember my time in A&E. 'Father' Ellis, the Consultant in A&E, was the first consultant to be appointed in this specialty in the United Kingdom. We performed a lot of minor surgery, I removed many sebaceous cysts and, on one occasion, did a skin graft.

During my time on the Eye Firm I watched Mr Black, the Consultant Eye Surgeon, expose a tear duct and remove a block. When it came to stitching up the wound on the face Mr George Black, a man of left wing sympathies, told the Senior Registrar that Mr Roberts would stitch up the wound and the Senior Registrar would assist me and at the end of that go and make me a cup of tea. The Senior Registrar was highly delighted!

Obstetric training was at Hyde Terrace and we had to live in the Hall adjacent to the Hospital, Croft Hall. We would be called out to see interesting cases in the middle of the night and assist. We were also called out on domiciliary visits and used a bicycle to get to various parts of Leeds. There were some worries about the safety of women around the rougher parts of Leeds at night but nothing seemed to happen.

We also had to go and live in at St James's. The residents mess was formerly a paediatric ward and the bathrooms, sinks, and toilets were all of paediatric size. It was extraordinary to see everyone getting on their knees to look into the mirrors when shaving. We slept at the end of an empty ward. At St James's after performing a delivery we had to get all the blood stained sheets and take them to a washing machine and wash them. At night we could not go to bed until all the washing was finished.

By the time we had started clinical studies several students had dropped out and we were numbered about sixty. The number of consultants was not large so we as individuals were known to most consultants as we would have attended their ward teachings or out patient clinics.

There was no doubt that one felt that the surgeons were the dominant force in the hospital which continued the Moynihan influence.

Surgery had made two major advances in the nineteenth century. In 1846 anaesthesia was introduced. This was in the form of a rag soaked in either ether or chloroform being placed by an inexperienced member of staff over the mouth of the patient. Infection however was the main problem and a major improvement occurred when Lister introduced carbolic sprays to the operating theatre. This was not the final answer however and Lister began to appreciate that infections came from dirty hands and dirty instruments. Surgeons wore frock coats at operation. These were so soaked in blood that they became rigid and could be leaned against the wall. Berkley Moynihan, when an Assistant Surgeon, remembers a newly appointed Surgeon using the coat of his predecessor.

Moynihan and his colleagues in Leeds decided upon a new approach. Instruments were boiled after use and Moynihan introduced the wearing of sterilised white coats and brought rubber gloves to Leeds from the United States. Moynihan developed the aseptic ritual of washing and dressing in sterilised white coats and gloves. Thus aseptic surgery was introduced with success. This permitted abdominal surgery which meant that a whole range of previously untreatable conditions could be cured by surgery. Moynihan wrote a book on abdominal surgery which became an international classic. His international reputation soared and he became the first provincial President of The Royal College of Surgeons. Subsequently he was ennobled. There were other great pioneers of surgery in Leeds and the reputation of the General Infirmary at Leeds for surgery soared. The operations were

11

performed in what is now the Instructional Block and Littlewood Hall. The Instructional Block has been little altered and the structure of the old anaesthetic rooms and operating theatres is still there.

The Leeds surgeons I knew had all worked for Moynihan and basked in his reflected glory. They were all charismatic individuals, which was important since before the introduction of antibiotics and blood transfusion it was important that the patients believed they were being operated on by one of the best surgeons around. Moynihan also made a fortune. His words to a newly appointed surgeon at the Infirmary were "the first thing you must do, my boy, is to make your first £100,000" a considerable sum in those days. In fact, my impression of the Infirmary as a student and houseman was that it was a temple devoted to the worship of Mammon. Many consultants married late, they sought a bride who was the only child of a rich industrialist.

The surgeons I knew were as follows. Digby Chamberlain was a tall dignified individual driven to and from the LGI by a chauffeur driven Rolls Royce. He was a skilled and precise surgeon, and his list was finished quite quickly. He did not believe that his patients ever needed blood transfusion. This was the era of the partial gastrectomy bonanza for the treatment of peptic ulcer. On every surgeon's list would be two or three partial gastrectomies and it was a considerable provider of private income. The result of this was that there were gastroenterology clinics to monitor the many side effects. Subsequently Professor Paul Fourman was appointed as a second Professor of Chemical Pathology and devoted his entire time to investigating the effects of partial gastrectomy on bone metabolism.

Michael Oldfield was a complete eccentric. He had a farm near Harewood and was a friend of the Princess Royal. He once did a ward round in full riding dress after riding to hounds. He was the son of a gynaecologist, Mr Carlton Oldfield. Mr Carlton Oldfield believed there were two types of women; those with backache; and those with severe backache. He made a fortune correcting retroverted uteri, a practice he abandoned when he had put his son, Michael, through Eton and medical school.

Perhaps George Armitage was the outstanding exhibitionist of them all. I remember going to an England and Australia test match at Headingley. The ground was packed. Len Hutton the idol of the crowd was opening the batting against his arch rival Ray Lindwall the fastest bowler in the world. Lindwall started a series of limbering up exercises and began to paw the ground ready for the signal to start when a tannoy sounded saying "would Mr Armitage the Leeds surgeon please telephone his secretary at the General Infirmary at Leeds". George then walked to the pavilion observed by a crowd of 30,000 people. On another occasion Mr Smiddy, a Senior Registrar to George Armitage was called to pick him up at Wakefield Station to take him to operate at Clayton Hospital. Smiddy walked onto the platform at Wakefield and was met by a porter who asked him which train he was expecting. When Smiddy told him he was waiting for the London train the porter said "I've worked here forty years lad and Leeds train has never stopped here once". Smiddy replied "it will today though" - and it did. George Armitage had been to see the Station Master at King's Cross and got his wish.

In George Armitage's consulting room he had a desk with the mounted hoof of a horse on it. When patients asked about this George would explain that this hoof was from his favourite stallion which was shot from underneath him as he led a cavalry charge at the Battle of the Somme. George would then limp round the room inferring that he had also received a war injury which was, in fact, due to a peace time motorcycle accident. At the time I was a House Surgeon we had a visiting Australian, Ken Brearley. When he was leaving George gave him a signed photograph of himself and took him to Lawnswood Cemetery for Brearley to photograph Moynihan's grave. He also gave Brearley a second hand biography of Moynihan which he had got by advertising. "Brearley" said George "I have read and re-read this book and it has brought tears to my eyes. In this book Jessop, a surgeon at the Infirmary, said to Moynihan "Moynihan you are the finest surgeon ever to have

13

worked at the General Infirmary at Leeds". "I wept because I very well remember Moynihan using the self same words when he spoke to me."

Of the other surgeons, Henry Shucksmith was a dynamic, irascible little man who pioneered vascular surgery. He would never go on holiday any further than Scarborough so that his secretary could contact him if he had a call to visit a private patient.

P J Moir and Johnny Latchmore were a much less demonstrative team. P J Moir became Dean of the Medical School, and Johnny Latchmore who was younger than any of the other general surgeons was a very pleasant and popular surgeon and played a major roll in the administration of the Infirmary in years to come. Mr Latchmoore was the last House Surgeon to work for Lord Moynihan before he retired.

Physicians of the Infirmary were a much quieter group with very different personalities. Professor Tunbridge was a full time University Professor and an expert in diabetes. He ran a large department with several senior lecturers with consultant status who performed most of the teaching. Each of the senior lecturers had a particular specialty and there was always the risk when one was newly appointed that his or her specialty would conflict with an already existing National Health Service consultant.

Rex Tattersall was an astute General Physician with a distinguished war record in Burma.

Otto Maxwell Telling was a Chest Physician and was the son of a previous Professor of Medicine. He did not believe in these new fangled antibiotics. I went on a ward round at St James's Hospital when he demonstrated that all his patients with chest infections had responded very well to sulphonamides. When he left the ward round the Registrar said "Sister! Let's do the proper ward round now and get out the other charts". These were produced and showed that in fact all of the patients were on penicillin.

John Towers was the son of a Leeds General practitioner and it is said that he got onto the staff by marrying the daughter of the Professor of Medicine. He was in essence a General Physician with an interest in cardiology. I remember one occasion when word passed round the Medical School that he had asked for leeches to be applied over the kidney of a patient with acute nephritis. He was a very pleasant man who was distinctly absent minded and I would work for him as his House Physician after qualification.

I saw little of Professor Hartfall, known as 'Happy Jack' which he distinctly was not. His main specialty was rheumatology.

Among the other interesting characters were the Dermatologists. The Senior Dermatologist was J T Ingram, something of a martinet, and with an international reputation as a Dermatologist. He went to an international meeting in Australia and was shown a lesion on the hand of a patient. Ingram made a diagnosis but was told that this disease did not occur in Australia. "It does now" said Ingram. It is not uncommon for liaisons between medical staff and nurses or sisters, but J T Ingram went one better and had an affair with the Matron, none other than Kathleen Raven who became Dame Kathleen Raven, Chief Nursing Officer of the Department of Health. Ingram was invited to the Chair in Dermatology at Newcastle and he and Dame Kathleen would meet at the weekend in their cottage in the Dales. A full length portrait of J T Ingram hangs in the Royal College of Physicians in London. The marriage of J T Ingram and Dame Kathleen was one between two very strong personalities and people wondered how they would get on. From what I heard afterwards the marriage can be summed up as the female of the species is more deadly than the male.

In the Department of Dermatology, Dr Hellier, was highly regarded and took a great interest in the histopathology of the skin.

The third Consultant was Steve Anning who has been the author of a distinguished book on the history of the Infirmary and Medicine in Leeds.

Hugh Garland, the Consultant Neurologist, was a larger than life character who dressed immaculately in morning suit with a fresh carnation in his lapel, and devoted his life to the pursuit of women. His appearance was commented on by the Princess Royal as she opened an eponymous ward. Garland's nickname was 'Pansy'. Garland would drive down through Headingley on a morning in an open topped Bentley car and if he saw an attractive women in a bus queue he would stop, extend his gloved hand, and invite her to take a lift. He would then proceed to meet his junior staff at the front door of the Infirmary and stand with his backside to the fire and inspect all the girls as they came into work. He occasionally sent a member of his junior staff to find out who a particularly attractive young girl was. He was a highly intelligent man who was a co-editor of a text book on General Medicine - *Garland and Phillips*. His main research interest was in islet tumours of the pancreas which secreted insulin and subjected the patient to hypoglycaemic attacks. One of the tests to diagnose this was a prolonged glucose tolerance test which lasted three hours. As students took all the phlebotomy samples, and samples were needed every half hour for three hours, it meant that students could take an entire morning performing this test. Complaints led to the creation of the first in-patient phlebotomy service in the United Kingdom.

I saw Garland mostly in the Out Patient Clinic. He was keen that his nurses were good looking; I actually saw him send for the Out Patient Sister and ask for the nurses to be changed as they were not pretty enough for him. He also used his teaching to vent his anger at the Professor of Medicine whom he disliked because of his University status. On one occasion there was a patient with an acoustic neuroma. After history and examination the students were asked what the diagnosis was. One student gave the correct diagnosis. He was then told he had made a diagnosis way beyond the capability of the Professor of Medicine.

On another occasion, a conversation with the students went as follows:

Garland: Who was Hughlings Jackson?

Students: The founder of British neurology, sir.

Garland: Correct! And where was he born?

Students: Don't know, sir.

Garland: Green Hammerton, near York. And what was the maiden name of his mother?

Students: Don't know, sir.

Garland: Garland!

He then walked in front of the students with a limp wrist showing his signet ring.

Garland: "This is the crest of the Yorkshire Branch of the Garland family". (In fact the name of Hughlings Jackson's mother was Hughlings.)

There is one character I have left until last - Philip Allison. He was a typical Moynihan protégée and set up a Thoracic Surgery Unit. He was idolised by the surgical staff, and many in the Infirmary. He was an adonis like figure, tall, bronzed, of athletic build and aquiline features. He had grey hair brushed down with hair cream and a diametrically straight parting. He wore an immaculately cut grey suit, a bow tie, and a coloured silk handkerchief drooped from his breast pocket. He, like many of his generation of consultants, kept his handkerchief up his shirt cuff; there was then a long tail of handkerchief dangling below his sleeve. Allison's tail was longer than anyone else's.

The University of Leeds announced that it was creating a whole time Chair in Surgery at the Infirmary and Allison was the unanimous choice of all staff in the Infirmary. Allison intimated that he was interested in the post but wished to retain his private practice. This was unacceptable to the University so there was an impasse. However, the University of Oxford heard of this and offered Allison a Chair in Surgery at Oxford with the opportunity to perform private practice. Allison, greatly honoured, departed. What seemed a disaster for the Infirmary was, in fact, a seminal moment. Allison's departure heralded the beginning of the end of the Moynihan era.

Geoffrey Wooler, became Head of the Thoracic Surgery Unit and took cardiac surgery into a completely new era.

However, surgery in the Infirmary was about to be rejuvenated with the appointment of Professor John Goligher. His first office was in the Instructural Block but when he arrived he had no clinical facilities. The Infirmary had not got round to allocating him any beds. I used to see him when I was a student ,walking up and down the main corridor of the Infirmary with little else to do. His reputation however rose rapidly. Patients with carcinoma of the rectum or the lower part of the sigmoid colon had their tumours removed and the remaining colon attached to a colostomy which was permanent. John Goligher, it was claimed, was adept at resecting tumours of the rectum and managing to anastamose colon to the remaining rectum. This avoided a colostomy. His students were subject to an extensive lecture program but he was not able to build up his department until the Martin Wing opened in 1961. In the new Martin Wing the Professorial Medical Unit had the top two floors (F and G) and the Professorial Surgical Unit the two floors below (D and E). He successfully created a department which produced several professors. The high reputation of the Infirmary for surgery was more than maintained.

In this chapter there have been many stories about the exuberant and larger than life behaviour of the Leeds Surgeons. In mitigation I should explain that patients expected this. With the advent of abdominal surgery people began to realise that they may need an operation at some stage in their life and would need to put aside money in the Building Society to meet this possibility. They then did expect value for money and this meant a visit by a distinguished looking surgeon in a gleaming Rolls Royce. There are two stories to follow which exemplify this.

On one occasion a patient who had abdominal symptoms asked for his General Practitioner to arrange a domiciliary visit. The General Practitioner arranged for a Physician to see the patient. The Physician, quiet in demeanour, took a full history, did a full clinical examination, and explained to the patient that he would need to come into hospital for further tests. The private patient was not impressed and asked the General Practitioner to arrange for a further opinion. George Armitage came to see the patient. There was a swish of gravel as the Rolls Royce entered the drive and George dismounted. He came to see the patient and told him that he was lucky in that he had just managed to catch George before he departed to Paris where he was about to read an important paper at a distinguished meeting of surgeons. There was then a ritual laying on of hands and George declared "I will get you in and put you right". The patient was highly delighted.

A second example refers an occasion when a surgeon, appointed consultant about the same time as I was, was asked to do a private domiciliary visit. He turned up to see the patient in his Mini and was asked by the patient if his Rolls Royce was in a garage being serviced! This whole pattern of behaviour was soon to change with the advent of the National Health Service. Finally I should point out that Professor Goligher drove a very modest second hand saloon.

I qualified in March 1957 and my first house job was as House Surgeon to Digby Chamberlain and Michael Oldfield. I had been on their firm and done two student locums.

The residents mess was really like an officers mess in the army. The Resident Medical Officer was a woman nicknamed 'Tweetie', the first woman doctor to ever hold a post of RMO at the General Infirmary at Leeds. She was head of the mess and on Sundays would sit at the head of a table and ask one of the house officers to carve. The common room and dining room and all the bed rooms were near the front door of the Infirmary. There was a party every night in the common room with music by Harry Belafonte,singing West Indian calypsos. We were busy but we had no worries about being fed and all our laundry was provided.

Most of the activity at the Infirmary was situated at the front door. A porter, who we all knew as 'Old Bill', was on duty most nights. He would give directions to visitors coming in through the front door and would answer the telephone which took all the general enquiries to the Infirmary. The way the front lodge connected to the residents was by a series of bells situated in each ward, usually over Bed 1. The porters pulled a handle and rang the bells according to a code. Mine was one bell followed by four bells. If this was repeated rapidly then this indicated an urgent call. Thus bells were ringing in the wards and on the corridors all over the hospital and it must have been hell for the patient situated just beneath the bell on a particular ward. The bells were replaced by electronic bleeps when I became a House Physician in September 1957. 'Old Bill' would be seen standing at the desk with a kettle, teapot, and teacup. Most people thought that the brown fluid coming out of the teapot was tea but it was, in fact, beer, and 'Old Bill' got livelier as the night went on. Bill answered all the queries about individual patient's well being. He had a sheet of paper from each ward with a list of patients and a code next to it, eg HOCS meant 'had operation condition satisfactory'. Bill used to embroider this quite a lot as he handled patient's queries.

Often parties in the mess got a bit out of order but when Arnold Tunstall, the Hospital Secretary (equivalent to a Chief Executive today) asked Bill what had gone on the night before he always stated that he had seen nothing, and usually that was the end of it.

In the evenings the bells stopped ringing and Bill had to find the House Officer by other means. This meant that he would come to the bedroom, enter, and wake the individual doctor. He always knew if there was a liaison between individual members of staff and knew exactly which bedroom to visit. One morning as he tottered home he took a short cut through a graveyard and fell in a newly dug grave and had to be rescued.

Life as a House Officer was relentless, eg a day on take the program would be as follows:

9.00 am meet Consultant at the front door and do a ward round.

10.00 am go to Out Patient Department.

12.30 check up on any acute admissions and take lunch.

1.30 pm proceed to Operating Theatre and assist Consultant and Senior Registrar until about six or seven in the evening.

7.00 pm again check acute admissions and get a quick meal.

11.00 pm visit acute admissions with Senior Registrar and prepare a theatre list and organise a theatre for about 1.00 - 2.00 am.

3.00-4.00 am operate.

4.00 am bacon, eggs and coffee with the anaesthetist and nursing staff, and then to bed at about 5.00 am.

9.00 am meet Consultant at the front door and go for a ward round.

Digby always arrived in a chauffeur driven Rolls Royce. The chauffer would alight and open the rear door for Digby to make a dignified entrance. We would follow him and we spoke if spoken to. We got to know what he was doing from what he told his private patients. On one occasion he said "I won't be in at the weekend. Prince Bernhard of the Netherlands is my house guest". In fact the consultant staff I worked for made a point of associating with the county nobility. Michael Oldfield was a particular friend of the Princess Royal at Harewood where he had a farm. While I was a House Surgeon he had a coming out party for his daughter, and the Duke of Kent attended. I was House Physician to Dr Towers and he was never available on a Saturday morning for he went partridge shooting with Lord Swinton.

Digby Chamberlain believed his patients never needed blood transfusion following an operation. What happened actually was that Sister Arnot would send for me after theatre and would hand me a cannula and I would put up a saline drip. She would then dismiss me and put up a blood transfusion when required, which was always taken down before the ward round next morning, and Digby was

21

never ever the wiser except that he was ever more convinced of his surgical prowess.

To a house officer the ward sisters were of enormous help. Sister Arnot on Ward 3, a Surgical Ward, had been a nurse in the army and she landed on the Normandy beaches on D-Day plus two or three. The ward was superbly run and spotlessly clean. All she needed to discipline an individual would be to give them a withering look. No consultants no matter how imperious their manner would ever cross Sister Arnot; they respected her enormously.

All the house surgeons had special jobs in Out Patients, for example, some were in charge of circumcision and others would perform minor operations. I had three tasks. One was to treat urethral strictures in middle aged to elderly men by dilating them with a metal bistuary; they were coated with glycerine and passed up the urethra in ever increasing sizes. Strictures were usually the results of gonorrhoea during World War I. The patients suffered quite some pain with this procedure but were always grateful afterwards. I also treated strawberry naevi in children by applying dry ice. Possibly my most pleasant task was the treatment of varicose veins in young women by the installation of some sclerosing fluid.

Michael Oldfield was an eccentric and not easy to work for. The theatre sessions were often a pantomime. He frequently accused me of being slow, and drove the nurses mad by hectoring them but they then got their own back by pouring very hot water into the bowls he washed his hands in. On one occasion he accused the Sister of being too slow and made me go to the other side of the table to thread needles. Mr Coates was a theatre technician and Michael would harass him. He usually asked him to move the operating table up and down most of the time. Michael Oldfield was tall and most of the staff assisting him had to stand on stools. On one occasion he said "take the table up Mr Coates". This was repeated several times until Michael was also standing on a stool and we all progressively rose higher and higher towards the ceiling of the operating theatre. On another occasion he decided to tell a joke and the theatres were cleared of all female staff and Michael then asked us if we knew what was the definition of a 'smile' in the fashion industry was. We did not know so he said it was the gap between the top of a lady's stocking and the bottom of her knicker leg. After a polite laugh the women were then allowed back in. Michael's specialty was in the repair of hare lips and cleft palates and my job was to hold a hook into either the ends of the cleft palate or the hare lip while Michael stitched up the cleft towards me. This was tedious and painful but every time I looked at the clock Michael saw me and told me off.

Private patients seemed to dominate however. My half day was on a Wednesday afternoon but that was the time that Digby Chamberlain admitted private patients into the Brotherton Wing and I had to clerk them and arrange the anaesthetist and theatre sessions. I never took a half day in surgery. Some times I was called out to give assistance in a nursing home. One of these was a former terrace house which had been converted into a private nursing home. The patient was carried from a bedroom almost vertically down the stairs into the front lounge. He had a block dissection of the glands of the neck (this operation is not carried out any more). There was a medium sized nursing home at the junction of Hyde Terrace and Clarendon Road. On one occasion George Armitage went with a private patient and asked for a bed. He was told that all the beds were full so he demanded to see Matron. He was told that Matron was on holiday so he took his patient into Matron's bedroom and placed her in Matron's bed!

One other responsibility I had as a House Surgeon was to look after private patients from one of the consultants at St James's. I have to point out that I did get a fee for assisting with private patients. I was not told this until the end of my house job and I had no record of any of the private patients so I got nothing. The experience I had as a House Surgeon will not be repeated at any time in the future.

Michael Oldfield died of a myocardial infarct in the late 50's, and Digby Chamberlain did not make the retirement age. George Armitage retired to Coxwold in North Yorkshire where he was known locally as the Major.

After my House Surgeon job I then moved to be a House Physician to Dr Johnny Towers. He was an avuncular, absent minded physician with an interest in cardiology. It was said to me that he was basically a high powered general practitioner. Life as a House Physician was very different. When I was on take I was on my own and if I needed another opinion I would seek the help of the RMO. Senior registrars and registrars did not do on-call. The experience I had was very wide ranging with all sorts of interesting clinical material. For example, I had two cases of cystic medial necrosis of the aorta which I diagnosed myself and organised for the Consultant Vascular Surgeon to come in and do what he could. I never rang my Senior Registrar or Consultant I just got on with the job. Old cases had a habit of being readmitted and so on days when I was not on-call I would have to manage acute admissions of old in-patients. I would find myself managing repeated bleeding from oesophageal and varices. I would insert a Sengstaken tube, an inflatable tube which has to be pushed down the oesophagus and blown up to compress the varices. The important thing however was that one would know the patient readmitted and management was not usually difficult.

There were two major events during my time as a House Physician. One was a period of dense fog and smog. Chronic obstructive airways disease was common in Leeds and many were admitted with cyanosis and CO_2 narcosis. The Nightingale wards were all full and beds were arranged head to tail from top to bottom of the wards leaving only Sister's desk. I would do a ward round with a laryngoscope and touch the vocal chord and provoke a paroxysm of coughing. If this did not work I would organise a bronchoscopy. The antibiotic of choice was then chloramphenicol which is no longer used because of the risk of aplastic anaemia.

The other major event was the epidemic of Asian flu. Many staff succumbed and the Brotherton Wing was full with members of staff. I remember that I remained well but one great tragedy was that it killed many patients with severe rheumatic heart disease.

The cardiac patients were very interesting and cardiac catheterisation had recently been introduced by a new consultant, Dr William Whittaker. I would often do a ward round with Geoffrey Wooler, the Cardiac Surgeon, look at the pressure tracings and make decisions about surgery. On one occasion Dr Towers took a six week holiday on a banana boat to the West Indies. I took his waiting list of about thirty and organised all investigations as an out-patient and referred them to Geoffrey Wooler. When Dr Towers returned he asked his secretary, Miss Anderton, if he could see the waiting list but was told that he had not got one. Dr Towers then proceeded to Out Patients and within a week the waiting list was restored to thirty. Dr Towers was also very absent minded. He would ring me up late in the evening and say that he had undertaken a domiciliary visit and would I get the patient in. He would then give me their name and address and ask me to organise a bed and transport. On several occasions I was informed by the ambulance service that the name of the street I had given was demolished some time ago. I would then ring up Dr Towers and get him out of bed. He then gave me the general practitioners telephone number and I had to ring him and get the correct address and arrange admission, which usually took place in the early hours of the morning.

I often received a lot of hassle from the private patients in the Brotherton Wing if Dr Towers had not been to see them. My predecessor, Dorothy Campbell, once arranged for all the private patients to come to the Out Patient clinic in wheelchairs who were to see Dr Towers as this was the only way she could get him to see them. The House Physician's job was one I enjoyed. What enhanced the experience was the fact that I was a Resident and never away and totally immersed in my job. In fact the whole year as a Resident was probably the most vivid and memorable year of my life. It was enhanced by the fact that one always had colleagues to discuss problems with and one could follow a patient from admission to discharge. The job satisfaction was enormous.

During the last two weeks of my internship I had a foretaste of my future career. A former Governor General of a Central African country was admitted to the General Ward with acute leukaemia. He refused to go into the Brotherton Wing. I met Arthur Bloom, Registrar in Haematology, who did a bone marrow and made the diagnosis. John Towers called in Dr J F Wilkinson to see him and advise on treatment.

Dr Wilkinson had a haematology practice which extended from North Wales to the North East of England, and he was a pioneer in the use of nitrogen mustard in haematological malignancy. I would meet Dr Wilkinson on many occasions during my future career and I attended his retirement dinner. Dr Wilkinson was a man of many parts. He was Consultant Haematologist at Manchester Royal Infirmary as a physician by training. His colleague in Manchester was Dr Martin Israels whose main interests were laboratory based. Dr Wilkinson claimed that workers making nitrogen mustard during World War I had neutropenia, that is a shortage of white cells. This led Dr Wilkinson to try nitrogen mustard in the treatment of lymphoma with success. This was during World War II and Dr Wilkinson has claimed that he was prevented from publishing his work due to the Official Secrets Act on the request of the Americans. They then proceeded to publish their results first.

He was a man of many parts. He was President of the National Aquarist Society which specialised in tropical fish, and was Medical Officer to Belle View zoo. He also acquired a fine collection of apothecary jars which were presented to the Thackray Museum in Leeds when he died. He was a great story teller and at his retirement dinner in Cheshire, where he lived, he went on speaking for hours. I remember the story of him in World War I chasing a German soldier along the harbour war in Zeebrugge with a cutlass. He finally pinned the German soldier to the wall with a cutlass at his throat but spared

26

him. He later met him at an international scientific meeting where the former soldier was making a scientific presentation.

This marks the end of my internship and I moved then directly to a post of Registrar in Pathology in the University Department of Pathology. So far I have described much of the early history of the Infirmary as it affected me but I have made no comment about the main subject of this book, namely, Haematology in Leeds.

In truth there was very little haematology. As a House Surgeon I very rarely requested any blood tests before surgery and perhaps a haemoglobin measurement after surgery if the patient had been bleeding severely at operation. The haemoglobin in those days was measured as a percentage of a standard and I was advised by the Senior Registrar to keep the haemoglobin above 80%. As a House Physician more haematological requests were made, mainly haemoglobins, white cell counts, and the Erythrocyte Sedimentation Rate. There seemed to be few haematological problems. Apart from the case of leukaemia, which I have just mentioned, the only haematological cases of interest were patients with hereditary spherocytosis who had been referred to the Infirmary for splenectomy, and pernicious anaemia.

Having said that I have to describe what was, at the time, the remarkable research leading to the treatment of pernicious anaemia. Pernicious anaemia was, until the 1920's an anaemia which progressed until the patient died. Also associated was an aggressive, degenerative, neurological condition known as subacute combined degeneration of the cord. Patients were progressively paralysed and demented.

Research began into this condition in 1918 when George Whipple conducted a series of dietary experiments on chronically bled dogs that led to the finding that the most effective dietary treatment was with raw liver. In 1925 Minot and Murphy were working at Harvard University on the liver treatment of pernicious anaemia and in 1926 published an account of their successful treatment of forty-five patients. The liver was homogenised and fed by intra gastric tube to the patient, but because of their B12 deficiency they were often very uncooperative and Minot and Murphy were often covered with liver homogenate and, on more than one occasion, were chased across the hospital campus by a pack of hungry dogs. The response to liver was confirmed by an increase in the reticulocyte count (immature red blood cells) and it was also noted that the megaloblastic marrow changes in pernicious anaemia were reversed. Thus there began an era when patients were fed raw liver to correct their anaemia and it was said that they were told to eat half-a-pound of raw liver per day. It was not long however before liver injections were prepared by pharmaceutical companies.

Then followed the classical work of William Castle in the same laboratories as Minot and Murphy. It was already known that patients with pernicious anaemia lacked the ability to secrete gastric juice. What Castle did was to feed a patient with pernicious anaemia 200 gms of rare beef a day for several days without improvement of the reticulocyte count, and then feed a liquid extracted from beef that had been pre-digested by Castle's gastric juice. This treatment with digested beef produced a full reticulocyte response. Further experiments followed and it was concluded that beef alone would not produce a response and neither would gastric juice. If the beef is digested with normal gastric juice it works very well. In 1929 Castle concluded that some unknown, but essential inter-reaction between beef muscle as an extrinsic (food factor) and normal gastric juice appeared to be required for the restoration of normal haemopoesis in a patient with pernicious anaemia. Castle termed the factor in normal beef to be the extrinsic factor, and the factor in normal gastric juice, but absent from the juice of a patient with pernicious anaemia, to be the intrinsic factor.

In 1948 scientists in the pharmaceutical industry in the United States and in Great Britain identified the principle in liver extracts active in pernicious anaemia as Vitamin B12 and Castle confirmed that this was in fact the extrinsic factor. Soon after this B12 was available for treatment of pernicious anaemia by injection. So important was this work deemed to be that Nobel prizes were awarded to some of the researchers in this remarkable story.

When patients with B12 deficiency are first treated with Vitamin B12 they soon begin to feel remarkably well and it was not very long before many patients were prescribed B12 when they attended their general practitioner or visit an Out Patient clinic feeling off colour. So before long the pharmaceutical industry was producing far more Vitamin B12 than could possibly be accounted for by the number of patients with pernicious anaemia. Dr Towers would often prescribe a liver injection called anahaemin for patients who had anaemia as a form of shot gun therapy. In the 1940's Lucy Wills who had gone to work in India from the Royal Free Hospital in London found that anaemic women in India responded to Marmite. It was soon found that the factor in Marmite was a folic acid and so this was a further factor for therapy of megaloblastic anaemia. This was the starting point of an enormous amount of work into nutritional anaemias involving the measurement of serum B12 and of serum folate and of B12 absorption studies but this is a topic to be enlarged in future chapters.

Serum B12 and folate levels were not yet available. The commonly used test was the Test Meal. The fasting patients had a drink and a piece of dried toast and then a gastric tube was inserted and sequential aspirations performed. These were tested for acid. An injection of histamine followed and if there was a histamine fast achlorhydria then this would be consistent with a diagnosis of pernicious anaemia but not diagnostic. In fact, many patients were found with a histamine fast aclorhydria and many were followed up and treated with regular drinks of hydrochloric acid before meals. Many of these patients were followed for years to see if they developed anaemia and indeed many of these were private patients. There followed however an enormous research interest, continuing for many years to come, on the nutritional anaemias.

CHAPTER 3

THREE GREAT PIONEERS

"No great man lived in vain. The history of the world is but the biography of great men." *-Thomas Carlyle*

As the Moynihan era faded three men began to take the Infirmary in a new direction; into the world of high technology. Geoffrey Wooler, Consultant Thoracic Surgeon; Frank Parsons, Pioneer in Renal Dialysis, and F William Spiers, Professor of Medical Physics. They were to put Leeds on the world map for their particular technologies.

Geoffrey Wooler was born in Leeds and educated at Leeds Grammar School, and Giggleswick. He qualified from Cambridge in 1937 and after the war he joined Tudor Edwards in London who had established a Thoracic Surgical Unit there. Philip Allison however had persuaded him to return to Leeds to join his newly formed Thoracic Unit.

In the early 1950's Denis Melrose, at the Hammersmith, was developing heart/lung machines, and of his three prototypes one came to Leeds. It is thought that Wooler financed part of the purchase price out of his own pocket.

After extensive work on greyhounds the first patient was operated on in February 1957. I remember a black and white BBC television series was produced called *"Your Life in Their Hands"*. The patient who was shown on television was in fact the first survivor of this procedure and remarked that as far as she was concerned the procedure was trouble free, but this was due to the fact that she had been unconscious for several days. The use of the heart/lung machine enabled numerous other techniques to develop, such as valve replacement. Observers came from all over the world to see and to learn the techniques involved with the heart/lung machine and Wooler received worldwide recognition for his work.

I got to know Geoffrey Wooler quite well as I used to take him on ward rounds to see patients of Johnny Towers who were being considered for operation. I found him to be extremely pleasant, courteous, and self-effacing - a complete contrast from the surgeons of a previous generation. Geoffrey Wooler also obtained cult hero status in the Leeds newspapers. Bob Appleyard the Yorkshire and England off spin bowler developed tuberculosis and Geoffrey Wooler removed the lesion from his lung and returned Appleyard to full health and the Yorkshire team. For many Yorkshire men at the time this was his finest achievement. Geoffrey Wooler once gave me a lift to the Test match in his Bentley. We drove through the main gate at Headingley, with a salute from the Commissioner, to a reserved car parking spot as befitted his status as a distinguished honorary member of the Yorkshire County Cricket Club.

Frank Parsons was a Leeds graduate and a surgical trainee who joined Leslie Pyrah's unit. Pyrah was a Pioneer Urological Surgeon but in contrast to many of his colleagues had an interest in research and led an MRC Unit investigating the causes of renal stone. His research unit subsequently became the MRC Unit for Mineral Metabolism situated in the Wellcome Wing.

Frank Parsons as part of his research programme went to Chicago to continue his research there and it was suggested that he visit the Peter Bent Brigham Hospital where a renal dialysis machine was being used. Frank Parsons was convinced of its value and he and Pyrah persuaded the Infirmary to buy the Brigham Kolff dialysis machine.

Apparently Frank Parsons was summoned by the MRC and told that *"our advisors say there is no place for an artificial kidney in*

British medicine". He was told by Sir Harold Hemsworth *"Parsons try it by all means but remember the country is against you"*. Notwithstanding, Parsons developed the first routine dialysis service in the United Kingdom. I remember the drama surrounding many of the case referrals from all over the country. Quite often a patient would arrive by helicopter on Woodhouse Moor and I remember visiting an air show at Leeds Bradford Airport when the whole show was stopped to allow a helicopter to land with a patient being referred for dialysis. Local newspapers made much of this. Many of the patients had renal failure following septic abortion but thanks to the renal dialysis team at Leeds a survival rate of nearly 100% was achieved.

I remember Frank when I was a House Surgeon. On acute take we would visit all the admissions to prepare an operation list to begin about 2.00 am. We would often bump into Frank wandering around the corridors of the Infirmary dragging on his pipe and gazing to the sky. He would often waylay us and spend about an hour discussing his most recent admission with the Senior Registrar. I did not know at the time that Frank could not consult anyone with experience, nor could he consult a text book or journal. Literally Frank had to start to treat patient with a blank sheet of paper.

F William Spiers was a Medical Physicist who was known throughout the world for his work on environmental radiation and he was a key seminal figure in the development of radiotherapy services in Yorkshire.

Radium, which was first identified and then separated in usable quantities from uranium ore by Marie Curie, is a radioactive substance which, in a series of nine steps, breaks down to form lead. In so doing it emits alpha particles, beta particles, and gamma rays. The alpha particles do not penetrate skin, the beta particles penetrate skin sufficient to cause damage, whereas gamma rays can penetrate deep tissues and so radium needs a protective shield. Radium was supplied in needles in platinum alloy tubes and these were generally implanted close to a tumour to treat it with gamma rays. To maximise dosage large radium sources were employed using lead shielding. In 1945 Leeds had a unit containing 5 gms of radium which was passed from a protective safe by a pneumatic tube to the beam head which emitted gamma rays for the treatment of a patient.

It was decided to treat patients with radium in the Infirmary in 1929 and a charity appeal was launched and raised £25,000 to purchase 0.5 gms of radium from the Belgian Congo. Other sources of radium were found which were much cheaper particularly from Canada. As a National (University) Centre Leeds received radium in needle form to supply the Radium Beam Unit.

In August 1934 the Infirmary Board of Governors received a letter from the British Empire Cancer Campaign (BECC) offering a grant towards the appointment of a physicist. The Board provisionally accepted the offer and in 1935 F W Spiers, then a demonstrator in physics at the University of Leeds, was appointed for a trial period of one year.

In the thirteenth annual report of the BECC, presented to the House of Lords specific reference was made to Dr Spiers who had begun to accumulate estimates of dosages of gamma rays and introduced new forms of protection. The Board of the Infirmary had no hesitation in keeping Dr Spiers on. This was an appointment of tremendous significance for Medical Physics and radiotherapy for he was to build a department of great international renown. Dr Spiers was on his own until 1944 because of the war. He continued to do research and produced significant numbers of papers. After the war Dr George Hay was appointed to help specifically with designing and development of apparatus. In 1945 a National Radiotherapy Centre had been established in the Infirmary. This was situated in the Infirmary in what is now the Department of Nuclear Medicine. This lead to the further development and expansion of the Medical Physics Department. Dr Spier's research continued unabated and his construction of a small high pressure ionised unit was to lead to the development of the whole body counter. In 1950 a radio isotope service was established and in this year the University Department of Medical Physics was established with Professor Spiers as Head of Department.

In 1951 the Röntgen award of the British Institute of Radiology was presented to Professor Spiers for his work on radiation absorption in bone.

Under the auspices of the Medical Research Council (MRC) and following the development of a high pressure ionising unit by Philip Birch, a Research Fellow in the Unit, the first whole body counter in Great Britain was built. This lead to the creation of an MRC Environmental Radiation Unit in the department with Professor Spiers as Director, and Philip Birch as Deputy. Space was now beginning to be a problem and Professor Spiers and Mr Leslie Pyrah the Urological Surgeon, both with MRC Units, approached the Wellcome Trust to make a generous grant towards the construction of a new building to house their research units. This bid was successful and, together with financial grants from the United Leeds Hospitals, the Wellcome Wing was built and opened in 1961.

In 1960 the Infirmary created a Department of Nuclear Medicine with Dr Clive Hayter, a Physician, in charge.

Meanwhile there was a desire to have more powerful units to produce gamma ray therapy. In the early 1950's radioactive cobalt was being produced in nuclear reactors. Radioactive cobalt emits gamma rays in great quantities and has the same radiation properties as the gamma rays from radium. Devices would have to be built with large protective shields.

In 1950 Professor Dainton, Professor of Physical Chemistry, informed Professor Spiers that he had received a provisional offer of a grant from the Rockefeller Foundation in the USA to buy a large radioactive source for research in radiation chemistry and asked if it would be possible to use it medically. Professor Spiers agreed but it would need in practice, a novel machine with a large movable shield. At this time the Regional Special Advisory Committee for Cancer Services of the Regional Health Authority was urgently considering where extra radiotherapy beds were to be placed. As treatment was becoming more complex and needed greater technical support Professor Spiers hoped that the beds could be placed on one site as the combined total of beds at Bradford Royal Infirmary and the Leeds General Infirmary was insufficient. Professor Spiers indicated that the presence of a Radiocobalt Unit, which because of the war time regulations could not be housed in the middle of a centre of population, was best placed at Cookridge Hospital, a convalescent hospital situated to the north of Leeds.

Plans for a Regional Radiotherapy Centre at Cookridge were fully accepted in 1954 as a result of collaboration between the Rockefeller Foundation providing the radiocobalt, the General Infirmary Endowment Fund building two heavily protected radiation rooms, and the University purchasing the laboratory equipment. The Regional Hospital Board built the connection to Cookridge Hospital, and the Yorkshire Council of the British Empire Campaign and the Rockefeller Foundation defrayed the cost of the elaborate dual purpose machine, and a radiocobalt source.This was a truly remarkable combination of effort that eventually resulted in an internationally acknowledged Centre for Radiation Chemistry and equally reputed Department of Radiotherapy.

Professor Dainton's work on the irradiation of chemical compounds was widely acclaimed and he was subsequently elected a Fellow of the Royal Society.

Plans for an appropriate building to house the Radiocobalt Unit were drawn up by the Regional Architect. Professor Spiers laid down principals of design for the Unit and local firms were brought in to manufacture it. Mr Allcock of the Hunslet Engine Company was appointed. He realised the principals of design and a large standard drill made by William Asquith Limited of Halifax was adapted to move the radio cobalt through the machine. The building was built by Messrs Wimpey & Company and a satisfactory conclusion was reached in May 1956 when the High Energy

Radiation Centre at Cookridge was opened by Princess Mary, the Princess Royal. This placed the Cookridge Centre at the forefront of radiotherapy in this country.

Thus, one can see the extremely important role of Professor Spiers in the development of medical physics in general, the creation of a unique environmental radiation unit in Leeds, and the development of high energy treatment of cancer in Yorkshire. Many research workers spend their lives actively seeking medical MRC grants and are proud to receive them.I doubt that Professor Spiers was ever without one. When he retired, the MRC actually commissioned him to work on the significance of radon gas in buildings.

I got to know Professor Spiers quite well and following a chat at a Nuclear Medicine Christmas party we developed a collaborative study in the dosimetry of radioactive phosphorous in patients with polycythaemia. Bill Spiers despite his many important achievements in Leeds was a very modest, self effacing character, and it was an honour to know him. He has a very special position on my Pantheon.

CHAPTER 4

THE UNIVERSITY DEPARTMENT OF PATHOLOGY

"Yours but to do or die, not to know the reason why."
Alfred Lord Tennyson - The Charge of the Life Brigade

This was the greeting I received when I entered the Department of Pathology from Dr Tommy Sutherland, Senior Lecturer in Pathology, with responsibility for Clinical Pathology.

I had long wanted a career in Scientific Medicine and I had been inspired by the teaching I had received in Pathology.

The University Department of Pathology was situated in the Algernon Firth Institute behind the Old Medical School. On the top floor was a magnificent museum, alas no more. The floor below held the University Department of Pathology, and below that the floor was shared between the University Department, Cancer Research, and Chemical Pathology. The University Department of Bacteriology occupied the ground floor and basement.

The Algernon Firth Institute was opened in 1933. The money for the building was raised by public subscription. The principle donor was Sir Algernon Firth, son of a Brighouse carpet manufacturer. He gave £25,000 through the Cancer Research campaign.

From the diaries of Matthew Stewart we know that a dinner was held with Viscount Lascelles and attended by the University Vice Chancellor, Sir Berkley Moynihan, and many other distinguished guests. Commitments for a further £30,000 were made, including Sir

Algernon Firth's donation, making a grand total of £55,000 the equivalent of £2,000,000 today. The five-storey building opened on the 25 April 1933 and is said to be inspired by a Dutch design in resemblance to Hilversum Town Hall. It is also probably the first building in the United Kingdom to use reinforced concrete. Lord Brotherton gave a further £1,000 for the equipping of the museum. Lord Brotherton had also endowed the Chair in Microbiology, which was first held by James Walter McCleod from 1922-1952.

There had been a Chair in Pathology since 1904 when the University was formed but the man who put Pathology on the map in Leeds was Matthew Stewart who first came to Leeds from Glasgow just before the outbreak of World War I. He was first based in Pathology in the Infirmary in the suite of rooms, which included the post mortem room, a series of laboratories, and offices, and overlooked the Nurses Home. There Professor Stewart ran a Clinical Pathology service for the Hospital embracing all disciplines including Microbiology, until Professor McCleod was appointed in 1922. Professor Stewart was renowned for his classification of peptic ulcers, his work in Industrial Medicine including asbestosis, and the founder editor of the Journal of Pathology and Bacteriology.

When Professor Stewart retired he was replaced by Rupert Willis an Australian with a considerable world reputation in the field of tumour classification and who was the Professor of Pathology when I was a medical student. He was a pleasant man who developed chronic pancreatitis and had to retire. His lectures were however incredibly boring as he read lectures from his student text book. He was replaced by Professor Charles Lumsden who had a reputation as an experimental neuropathologist and who devoted his life's work to finding the cause of multiple sclerosis.

The University Department of Cancer Research was again very highly regarded. They developed a technique for evaluating the carcinogenicity of the dyes used in the Bradford wool industry by inserting a wax pellet impregnated with the dye into the bladders of rats. The Head of Department was Professor P D Massey who retired in 1953.

Chemical Pathology was a relatively small Department and at this stage had not undergone the tremendous expansion which was to follow in the 1960's. G H Lathe who was known for his important work on the breakdown of bilirubin in the liver was appointed Professor. Chemical Pathology moved from the Algernon Firth Institute to the basement of the Martin Wing when it was completed in 1961.

The University Department of Microbiology was again a Department of great renown. Professor McCleod had classified the diphtheria organisms and was, in fact, Dean of the Medical School when I arrived. He was followed by Professor C L Oakley who was a leading role authority on the anaerobic clostridium organisms. We, as students, were amazed by Professor Oakley and his unbelievable knowledge of a wide range of subjects. Apart from Microbiology, he spoke fluent Chinese, was an expert on medieval church architecture, and science fiction. He had a comprehensive knowledge of the topography of streets in London, and was very interested in natural history. During the war he was arrested when found shining a torch into horse troughs at night to examine the local flora and fauna. He was accused of signalling to enemy aircraft. Professor Oakley was elected a Fellow of the Royal Society during his time in Leeds and eventually was awarded the CBE.

As a House Physician I went to see Professor Lumsden for career advice in Pathology and he offered me a post of Registrar which I was very pleased to take. There was a Training Scheme in the Yorkshire region which involved spending one year at the Infirmary, one year at St James's and a further year or two in a smaller District General Hospital. This did not appeal to me.

So, the day after I finished my House Physician job I joined the University Department of Pathology as a National Health Service Registrar. The Hospital service for Pathology was situated in the Infirmary buildings from which Professor Matthew Stewart ran Clinical Pathology before and after World War I. In fact, it was my impression, that apart from losing Microbiology very little had changed since then. I was shown round the laboratory suite. The first room we entered was Histology processing. There was a bench around the walls and in the centre of the laboratory a space. This was occupied by a motorbike belonging to the Chief Technician, George Corry, which had been taken to bits and was being patiently restored. No one seemed to take any exception to this whatsoever. The next laboratory was Clinical Pathology where there was a central table upon which blood counts were performed. Other tests were performed however, such as, urine microscopy, the examination of cerebral spinal fluids, the calculation of basal metabolic rate, and examination of hot stools for amoeba. It was staffed by a Chief Technician and about eight technicians, all of whom were young, very attractive, women.

The Registrar in the laboratory was the Registrar in Haematology, Dr Arthur Bloom, who was to become a lifelong friend. He had been in post for two or three months and his post had been created at the same time as mine. His post was, however, that of Registrar in Haematology. When he arrived in the laboratory on his first day he asked to be directed to the Department of Haematology and was told *"You are it!"* Dr Tommy Sutherland, in charge of the laboratory, was known as Senior Lecturer in Clinical Pathology.

In fact Leeds, apart from the Hammersmith Hospital, was no different to any other hospital in the country as independent laboratory based Departments of Haematology did not exist. Virtually all research in Haematology at the time was provided by Academic Departments of Medicine.

The staffing structure of the Department of Pathology was very good with specialists in virtually all disciplines. For the most part they were situated close to their clinical colleagues. For example, Dr Keith Anderson who was a Urological Pathologist, had a laboratory in the Department of Urology. Dr Colin Woods, the Bone Pathologist, was situated in the Department of Orthopaedics, and Dr Denis Harriman, the Neuropathologist, ran a Neuropathology

Department in the basement of the Brotherton Wing close to the Neuropathology operating theatres.

The medical staff however were united in their hatred of the Professor. Professor Lumsden exhibited a degree of parsimony far in excess of that which might be expected from an Aberdonian Scot and reigned with a degree of absolutism not seen since the Court of the French King Louis XIV. There were no staff meetings, no scientific meetings, case discussions, or slide meetings. Moreover, the Professor prepared all the work rosters which could not be varied. The Professor also insisted that all histopathology reports had to be written in longhand as it was not legal to dictate them, so he introduced a bound book in which reports had to be written by hand. On the cover there was a line pattern of butterflies and it was known as the 'dreaded butterfly book'. The secretary then typed reports from this book. Dictaphones were forbidden.

The Professor had research laboratories at Clarendon Road where he investigated the pathogenesis of multiple sclerosis and was funded by the Multiple Sclerosis Research Fund. No members of the routine medical staff ever visited these laboratories. The Professor referred to the medical staff of his Department not involved with him in his research as 'routine jacks'.

The first two weeks in my new post were spent with Dr C J E Wright reporting histopathology sections in the Algernon Firth Institute. Dr Wright's office was remarkably tidy and constructed of polished wood throughout. The atmosphere was monastic and there was no clinical contact. After this two weeks, I fled to the laboratories in the Infirmary and made a place for myself in the Clinical Pathology laboratories, enjoying the hustle and bustle of a busy Department situated in the middle of a busy hospital. I have never since believed that laboratories should be situated out with the main clinical areas. My job soon consisted of processing and reporting centrifuge specimens of urine, and calculating basal metabolic rates. These were performed by a technician from Chemical Pathology, called Bob, who visited the patients with a spirometer that measured air uptake of air as the patient breathed. Bob was a man of few words who rammed the mask on the patient's mouth then watched as a jagged line appeared on a rotating drum on the machine. The piece of paper was labelled and sent to me. I then drew a line through the tracing and calculated the slope and from that the basal metabolic rate. Whether any of this had any value whatsoever in the diagnosis of thyrotoxicosis I have my grave doubts.

I was also taught how to do bone marrow aspirations by Dr Sutherland. He showed me how to identify the cells of the bone marrow and then I was off on my own. Thus began, and ended, my post graduate education in Haematology. The rest I taught myself and picked up things as I went along. There was a problem however in reporting films and bone marrows - there was only one microscope in the Department which had lenses sufficiently good enough. Often during the reporting of a bone marrow the technicians would ask if they could have the microscope back to do some differential white counts. After innumerable protests Professor Lumsden gave the Department an old second hand microscope of reasonable quality. The microscopy bench was like many others,it bore many circular stains from tea and coffee mugs and the edges shewed many cigarette burns.Health and Safety was but a twinkle in some Beaurocrat's eye.

Arthur and I worked well together. Arthur worked in Haematology only, but my program was a three monthly one - one month in autopsies, two weeks reporting in Histology, and six weeks in Haematology. It was then that Mr Wooler's team approached Arthur and asked him to investigate the coagulation changes associated with the heart lung machine. There were recent developments in the field of coagulation emanating from the University Department in Oxford run by Biggs and Macfarlane. A recent test described by Douglas and Biggs was the thromboplastin generation time which facilitated the diagnosis of haemophilia, or Factor VIII. Deficiency. Thromboplastin is the product of a reaction of various clotting factors which finally convert fibrinogen to fibrin, the main constituents of a blood clot. The clotting factors can be divided into those which are consumed in clotting and those which are not, and are to be found in a serum. Serum is the fluid left when the blood has clotted. The stable factors can be removed from plasma by absorption with aluminium hydroxide. Therefore absorbed plasma contains Factor V and Factor VIII of which Factor VIII is deficient in haemophilia. Serum contains Factor VII and Factor IX. Factor IX deficiency is the cause of a haemophilia-like syndrome called Christmas disease, after Stephen Christmas the first patient to be discovered with this disorder. To do the test absorbed citrated plasma is mixed with aged serum and platelets. The mixture is then recalcified and thromboplastin generation begins and a stop clock is started. At minute intervals a sample is taken from this and this is added, with calcium, to citrated plasma and the clotting time recorded. If there is a factor deficiency then these clotting times are very slow.This sequence is repeated with patient's plasma,patient's serum,and a control mixture of normal reagents. This test requires considerable manual dexterity and speed as both hands are used in pipetting and stopping and starting stop watches. This test can be further developed to perform Factor assays. Arthur became an expert in this test and agreed to the research proposal.

Professor Lumsden was then asked to provide a deep freeze, two stop watches, and 0.1 ml pipettes. Arthur was given an old ice cream tub with dry ice for a deep freeze and told to calibrate Pasteur pipettes by sucking up a given weight of mercury. To make a Pasteur pipette a length of glass tubing is heated on a Bunsen burner until it begins to melt. It is then pulled out into a thin tube in the middle which is divided when cold. A rubber bulb is attached to the other end and this is used to suck up and blow out units of fluid. This pipette had a widespread use in scientific laboratories, particularly in Serology and Blood Transfusion. I very well remember Arthur chasing drops of mercury all over the bench to calibrate these pipettes. An old stop clock was found in the laboratory and Tommy Sutherland bought Arthur a stop watch out of his own pocket, and so Arthur began his project. There were problems because blood samples were often contaminated with heparin. However the investigation was a success and was duly published.

Arthur then began to look for another job and eventually was appointed Lecturer in Haematology at the University of Wales in Cardiff. He married Jane, a Neuropathology technician, and departed to undertake a very distinguished career in Haematology and was Director of the Haemophilia Unit in Wales, and became one of the first persons to describe the structure of the Factor VIII molecule. I remember very well meeting Arthur to undertake a Joint Committee on Higher Medical Training (JCHMT) visit in London. As we visited a Haemophilia Centre it was obvious that Arthur was very distressed when he met haemophiliacs who had developed Aids. He had known many of his patients for years, and were virtually friends when they had developed Aids from blood products which he had often personally administered. The development of Aids in patients with haemophilia was a great tragedy with widespread repercussions among patients, relatives and the medical staff who had looked after them.

The principal function of the Department was to provide a blood count service as it had done for many years. The measurement of haemoglobin was a cause for concern in the years following World War II and was a source of interest for the Medical Research Council who devised a colour comparison machine which was still in use in Blood Transfusion, at the Hammersmith Hospital, when I was there in 1962.[page 119]. Haemoglobin was measured by a range of methods of visual comparison by the observer. One day cleaning out cupboards in the Department I came across a range of haemoglobinmeters and these, with other systems collected along the way, are shown [pages 116 to 129 with an explanation of techniques used. An arbitrary standard of 100% was chosen as 14.6 gms per 100 ml of blood by the MRC. For many years, in clinical practice, the haemoglobin was expressed as a percentage of a standard and when this was changed to grams per deciliter introduction was slow and arduous.

Before automation the MCHC was the most accurate of the absolute values because both haemoglobin and PCV are very reproducible. The absolute value, mean cell volume (MCV), was

44

calculated from the ratio of the PCV over the red cell count but, unfortunately, visual red cell counts were too inaccurate for this to be of any value and it was not unknown for bone marrow tests for megaloblastic anaemia to be performed on a patient with a falsely raised MCV. This was completely reversed when automated cell counters became available and then the MCV was a very valuable measurement indeed.For a blood count to be provided a haematology technician had to visit the patient. The most common request was for a haemoglobin only but very often a white cell count and film might be required and, less frequently, a platelet count. On visiting the patient the technician would jab the patient's finger with a Hagedorn needle. This was triangular in section with a sharp point. It was mounted on a cork which fitted onto a tube containing spirit to sterilise the needle between patients. The bottom of the tube was usually covered in droplets of inspissated blood. The degree of sterilisation was doubtful; I remember one patient who was bright yellow with Hepatitis B. On the ward, 100 days later, there were three more bright yellow patients. In those days nobody seemed to bother very much. Samples of blood were sucked by mouth pipettes from the drop of blood on the finger. In the case of haemoglobin the blood was sucked to the mark in a straight pipette and then blown into a measured volume of diluent in a tube. A film could be made from the blood on the end of the pipette. For cell counts blood was sucked to a mark on a bulb pipette. A diluent was then sucked in to a mark above the bulb and then the mixture of diluent and blood was mixed by vigorously rotating the tube between the hands. After blowing out the contents of the stem the mixed sample was applied to a counting chamber.

If an erythrocyte sedimentation rate (ESR) was requested then a venous sample had to be taken into a citrated tube and the blood sucked up a mark in a graduated Westergren ESR tube in the laboratory and the blood column read after one hour. This was preferred to the Wintrobe method.

The technicians were very skilled and fast. The samples were brought back to the laboratory and placed, with the request card, on a large table. The haemoglobin samples were read in a colorimeter in a spectrophotometer with a fixed wave length called an EEL colorimeter. The cell counts were performed in counting chambers. These were wells situated on a deep glass slide with a measured grid on them. A colour slip was put in place and the well filled with diluted blood by capillary action. The cells were counted using a microscope over a fixed area which represented a volume of blood. A quick calculation done in the technician's head produced a cell count. Platelet counts were performed by this technique when requested. Red cell counts could also be performed this way but were often inaccurate because of the large number of red cells in the chamber to be counted. Haematocrit or packed cell volume (PCV) were performed using the Wintrobe tube but before centrifugation took place the blood was allowed to settle and Wintrobe ESR read. The Westergren and Wintrobe ESR's were not strictly comparable and subsequently the Wintrobe ESR was abandoned. The results of all the counts were written on the back of the request card then this was typed by the departmental secretary. Medical staff at the end of the day would sign the typed copy. Occasionally errors of typing would slip through. On one day the secretary got the initials of a Mr Fred Tucker transposed and this was missed. Over 200 counts per day were performed. It should be noted that virtually every procedure performed at this time contravened the present Health & Safety Rules and to perform any of this mouth pipetting in this day and age would lead to instant dismissal.

Despite the speed and dexterity of the technicians the system was at about its limit because of the need to visit the patient. There followed a virtual revolution in blood counting around 1960 when an anticoagulant sequestrine, that is, ethylene diamine tetra-acetic acid (EDTA) was introduced. As long as a film was made from the sample within one hour the cell morphology was unaltered. It is a dry anticoagulant so there is no volume change associated with it. It meant that venous samples could be taken on the wards and sent to the laboratory. Technicians' visits to the ward were now not necessary except for some specialised tests. There were not however the funds to make immediate changes. All the syringes were glass and needles metal and needed sharpening. At the end of the procedure syringes and needles had to be washed and sterilised by heat and this became the function of the laboratory washer-up and there were quite simply not enough syringes. The washer-up also had to empty and wash the sample bottles which were 5 ml glass bottles which were screw capped and known as Bijou bottles. They had to be washed, dried, a volume of liquid sequestrine added, dried again, and then labelled. There were therefore still problems in providing an expanding service. All coagulation studies not only had glass syringes and metal needles to be washed and sterilised, but also they had to be coated in silicone and then dried again so that the glass syringes would not activate the coagulation process. Finally the next revolution was the plastic one which led to the disposable age. Thus there was a watershed in the evolution of haematology. The service moved from a ward based collection service to a laboratory based diagnostic service. The main changes leading to this were the introduction of sequestrine and the disposable plastic revolution. Detergents were just being introduced into general use for cleaning purposes and were a boon for the washing up lady.

About this time Professor Lumsden decided to move the Histopathology service to the Algernon Firth Institute and this included Dr Sutherland and the other lecturers. The large central table was replaced by island benches which were at first unpopular but as the Histopathology laboratory was emptied it doubled the space available for haematology. It also meant that I, a Registrar with eighteen months experience, was Acting Consultant Haematologist to the General Infirmary at Leeds, a major Teaching Hospital. It became obvious to Professor Lumsden that with a major service controlled by a junior doctor he needed some element of control. This lead to the creation of a Senior Chief's post and this was filled by Charlie Buchan a Chief Technician from King's College Hospital in London.

None of the technicians in the Department had any basic training apart from that received in the Department. They were all very pleasant young women and very good at what they did but with little scientific training. Dave Greensmith was in charge of them and he led them very well but he again had no scientific background. Professor Lumsden had a policy of appointing technicians who were shapely and good looking young women and who would cheer the patients up when they visited them. I think I can say that I disagreed with virtually everything Professor Lumsden did but not this. I think it is quite true that the technicians who work in the Haematology laboratories today are as shapely, good looking, and pleasant as ever but nevertheless very skilled and knowledgeable. When Charlie arrived he had an enormous task to change the culture of the Department.

Charlie and his wife Celia were both fully qualified as technicians and held the Fellowship of the Institute of Medical Laboratory Technicians and had worked in a Teaching Hospital Department of distinction. Charlie and Celia were married at King's. Celia left and went to work with a Dr Ackroyd at St Mary's hospital and had assisted Ackroyd in some of the first work to be published on platelet antibodies.

Their move to Leeds in the early 1960's was a time of great change in the United Kingdom. Dress was becoming brighter and it was the beginning of the rock and roll age with the Beatles and the Rolling Stones. It was also a time of significant immigration into the United Kingdom, particularly from the West Indies. The first immigrants were mainly male and moved to specific areas in major towns which, in Leeds, was the suburb of Chapeltown. With the pattern of immigration of young males these areas soon began to be associated with drug problems, marijuana, prostitution, pimping and knife and gun crime. To help Charlie and Celia and their two young children find permanent accommodation the University of Leeds offered them some temporaey University accommodation bang in the middle of Chapeltown. Fortunately their salaries were adequate for them to get a mortgage and move to a house in Cookridge in a short space of time.

Charlie and Celia were London Eastenders and Leeds presented a very different world. One day Charlie mounted a double decker bus as it slowed down round a corner only to be met by a big breasted Yorkshire conductress who got hold of him by his coat lapels and ejected him at the next stop. Charlie noted that in London workers went to work on public transport in decent clothes and changed when they got to work into their working clothes. In Leeds all went to work in their factory overalls and deposited their dirt all over the bus seats. Still at the time Senior Chief Technicians were relatively well paid, much better than now, and Charlie and Celia soon settled down and Charlie even became a Leeds United football supporter and went to the matches with me.

There were some promising individuals among the technicians, Kathy Holmes as she then was, was the first Infirmary Medical Laboratory Scientific Officer (MLSO) to go to Technical College and pass the Fellowship of the Institute of Laboratory Medical Scientists. From the beginning however Charlie had bother with women. Mary Quant had just introduced the mini-skirt and one Monday morning all the girls, a very good looking and shapely bunch, turned up in mini-skirts and departed to the Wards. As they bent down to take samples they revealed an area of bare thigh covered only by a black suspender. This devastated the middle aged and elderly male patients and ECG monitors went wild ;anxious Sisters telephoned Charlie asking him to withdraw the technicians at once. I remember Charlie sitting down with the girls with a serious face telling them that if they wished to wear mini-skirts then they would have to wear tights at the same time.

One young woman called Val gave Charlie a torrid time. On one occasion she used the balance to weigh out some Geimsa stain used for staining blood films but she omitted to clean the balance afterwards. Charlie was the next person to use the balance and mopped his fevered brow several times. When he had finished this task his face was bright purple. He departed the Laboratory enraged and went on the hunt to find Val but she had beaten a hasty retreat.

As Charlie's career progressed he found himself in charge of more and more women, mainly phlebotomists. Hugh Garland was known for identifying the hypoglycaemic syndrome associated with islet tumours of the pancreas and he had devised a prolonged Glucose Tolerance Test which lasted three hours. At first this was performed by medical students but they complained about being occupied doing phlebotomy for three hours on a morning which was a complete waste of time. As a result two phlebotomists were appointed to take care of the various Chemical Pathology balance tests including Glucose Tolerance Tests which entailed an early morning start as the patients were starving. Also the Surgeons created two more posts to make sure that blood samples were taken in time for Theatre. The Department of Chemical Pathology looked after the first phlebotomists to be appointed but as the number increased passed the job of supervision over to Charlie. Charlie and I were subsequently approached by Sister McClutcheon to start an Out Patient Phlebotomy Service and this we eventually did so Charlie had more and more women to look after and he became a sort of Father Confessor and a Marriage Guidance Councillor to the mainly middle aged women.

Early in the days of the phlebotomy service we appreciated that the bloods from two patients in a ward had been mixed up because one of the patients on the ward had leukaemia and had been given another patients name. This was a mistake of the medical staff. I appreciated however that there was no certain way phlebotomists, who did not know the patients on the wards, could identify them positively. I therefore wrote to the Committee of Physicians to explore whether or not identification labels could be created for patients. This was turned down flat. One of the Physicians had been a Japanese prisoner of war and there were memories of inmates of concentration camps having identity numbers tattooed on their wrists. The proposal did not see the light of days for some years to come as the result of these war time experiences.

The last three phlebotomists Charlie had to manage were Pauline, Rita, and Jean, known by everyone as "Charlie's Angels" but angels they were not! Jean was an ex Tiller girl dancer and one morning her husband, whom she suspected of infidelity, awakened to hear two clicks. As he opened his eyes he found himself looking down the barrel of a double barrelled shot gun; the two clicks were made by

Jean as she cocked the triggers. The girls, however, were excellent workers and never made any mistakes.

The major field of development in Haematology at the time was in Coagulation and Charlie spent a considerable time developing tests in this field. The basic screening test in early days was the whole blood clotting time in which three glass tubes were filled with 1 ml of blood and incubated in a thermos flask on the Ward and the clotting time of three tubes measured by a stop watch. This test was basically no different to that performed by Thackrah in the early 19th century and was not of much value in detecting coagulation abnormalities. The other principal test was the bleeding time which at first was performed by pricking the ear lobe and blotting the ear at regular intervals with filter paper until the bleeding stopped and the time recorded. This was later modified to involve a standard cut on the forearm of the patient with a sphygmomanometer inflated to 40 mm of mercury. The screening tests for coagulation abnormalities used in the early 1960's was the Kaolin Cephalin Clotting Time. This is basically a modified plasma clotting time.   If blood is collected into anticoagulant citrate and the red cells centrifuged and removed plasma is left. This will clot if calcium is added but the results are very variable.   The tests can be greatly improved by adding kaolin to standardise the activation by glass contact and by adding a platelet substrate known as cephalin (a chloroform extract of the thromboplastin used in the prothrombin time). This is known as the Kaolin Cephalin Clotting Time.

The major test for coagulation disorders was the Thromboplastin Generation Time developed at Oxford in the Department of Douglas and Biggs who were leading authorities in this field. This test had been first performed by Arthur Bloom as part of his studies into the clotting changes associated with open heart surgery. This has been already described. There are however very many problems with this test and one mistake I made more than once was to drop a glass pipette into a glass tube too heavily and crack it thus letting bath water into the sample and making all samples incoaguable. The whole test had to be abandoned and restarted with fresh reagents and we would be occupied late into the evening getting this ttest to work.

Performance of the Thromboplastin Generation Time was a rite of passage for all trainee haematologists and its performance for some time was a critical part of the final examination for membership of the Royal College of Pathologists in Haematology. Without a skilled performance in a Thromboplastin Generation Time one could not pass the MRCPath final and therefore one could not become a Consultant Haematologist.

No mention has been made of the Hospital Blood Transfusion Service. When Blood Transfusion Services were introduced to hospitals the University of Leeds in its infinite wisdom declared that as Blood Transfusion was not an academic subject and it had no interest in the matter whatsoever. The Blood Transfusion Service therefore was provided by the Laboratory in the Hospital for Women. The Hospital for Women was situated immediately next to the Algernon Firth Institute and was subsequently knocked down to build the Clarendon Wing. The United Leeds Hospitals comprised the General Infirmary at Leeds, plus the Hospital for Women, and the Maternity Hospital situated in Hyde Terrace. Both the Hospital for Women and the Maternity Hospital had independent laboratories, both providing a comprehensive clinical pathological service.

The Pathologist at the Hospital for Women was a Dr Lissimore who was basically a Histopathologist who reported gynaecological pathology. There was no Gynaecological pathology in the main Department of Pathology in the Algernon Firth Institute. We used to visit Dr Lissimore for teaching during our training in Gynaecology as he would regale us with innumerable stories about the present Consultant staff who were his fellow students in the Medical School. He was joined later by Dr Alan Ambery-Smith a dapper, small man who wore a monocle and bow tie. Dr Ambery-Smith was an excellent Obstetrician who taught me as a student and was an expert with forceps, but because of the general economic situation there were no Consultant posts,certainly he failed to get one post because he was considered too good for that Hospital. He was however seconded to the Hospital for Women for training with Dr Lissimore as a Gynaecology Histopathologist, and also took charge of Blood Transfusion. Dr Ambery-Smith also did all the private practice pathology on the Brotherton Wing and did a visit every morning to take samples from patients. The laboratory at the Maternity Hospital was in the charge of Dr Kohler and again this was run well but I had little contact with it. I did, however, get to know the technicians in both laboratories, they were all very good and subsequently with future reorganisations became part of the laboratory staff at the Infirmary.

The Blood Transfusion Service, as far as the laboratory was concerned, was an excellent one but there was the problem of transporting blood to the Infirmary. This had to be done out of doors and the blood was transported by a Porter called Ian Leaf who was known to all as "tealeaf". He would wheel his barrow backwards and forwards to the Infirmary most of the day and when it was raining would wear a sou'wester and oilskins and closely resembled the fisherman with a cod over his back on the label of a bottle of cod liver oil. The blood was delivered to the Infirmary and to the wards in all weathers where it was put in the general domestic refrigerator with the butter, the sugar, and the milk. As soon as the blood left the laboratory it was out of any temperature control whatsoever and these conditions were grossly unsatisfactory.

The laboratories at St James's were also a complete Clinical Pathological Service with no specialisation. The Head of Department was Dr Bill Goldie, a Clinical Pathologist of renown, known to be one of the best Clinical Pathologists in the country. Dr Goldie also set up the first Haemophilia Service in Leeds. Overall, it was said that the hospital service at St James's at that time was better than the Infirmary's with its many specialists.

The clinical haematology service was performed by Dr J R Fountain who was an Edinburgh Graduate with an MD with gold medal for work on 6-mercaptopurine therapy in experimental mice.with leukaemia. He was Clinical Tutor in the University Department of Medicine. He saw the haematology referrals and ran the Friday Clinic which was known as the Isotope Clinic from the days when Radiotherapy was situated in the Infirmary. Most of the cases of polycythaemia vera in the Region were referred here for treatment with radioactive phosphorous and Jim was able to publish a large series of results of treatment in patients attending the Infirmary. He was usually assisted in the Clinic by a Research Fellow from the University Department of Medicine. At this time the condition of essential thrombocythaemia was not known and some patients were referred to the Clinic with very high platelet counts and haemorrhagic symptoms. These were given the name of haemorrhagic thrombocythaemia by Jim and were written up as a new entity. They were treated successfully with radioactive phosphorous. The paper was a joint one with the Research Fellow and Dr Sutherland who reported all the bone marrows. I remember one morning Dr Sutherland, the meekest man I have ever known, appeared to be in a tremendous rage with a very red face, slammed the door of his room and departed to see Professor Lumsden. Apparently Dr Fountain had written up his paper with no names or acknowledgements to any of his collaborators. There was obviously an enormous row, of which I know little, but it meant the end of Jim's academic career and he ultimately was appointed as a Consultant Physician to Keighley before the Airedale & Wharfedale Hospital was built. He did however return to do the Friday Clinic.

What made matters worse was the fact that Jim was a member of the first Medical Research Council Working Party for Adult Leukaemia which had instituted a trial of 6-mercaptopurine versus 6-mercaptopurine plus prednisolone in acute leukaemia. Many young patients were referred to the Professorial Medical Wards from all over West Yorkshire and, of course, failed to respond and died. The Professorial Medical Wards which were currently being transferred to the newly opened Martin Wing were a dreadful sight and two new Sisters, Sister Freda Ellis and Sister Rita Cox, were appalled. They helped many patients out of this world with appropriate mixtures of opiates. The Sisters declared that never again would they allow an

MRC Trial on their wards. A grave omen for the future of Haematology in Leeds.

Also appointed about this time was Clive Hayter as Consultant in Nuclear Medicine. Clive was basically a Physician by training and his new Department was situated in the old Radiotherapy Department of the Infirmary. It was the function of the Department to provide all tests and therapy involving radio isotopes and this included tests in Haematology, Chemical Pathology, and Radiology. This would cause conflicts in years to come leading to several personal tragedies. When Clive was referred patients for blood volumes and was able to diagnose polycythaemia vera he would often keep the patients and refer them to his own Clinic.

Meanwhile Professor Lumsden, no doubt under pressure, decided to appoint a Senior Lecturer in Haematology with Consultant status. This was filled by John MacIver who had been working in the University College of the West Indies. His main interests were in the haemoglobinopathies and megaloblastic anaemia and his work did not clash with mine. He kept Charlie busy developing haemoglobin electrophoresis and the Figlu Test. This test measures the accumulation of formiminoglutamic acid (Figlu) in the urine in B12 deficiency. B12 deficiency causes a metabolic block in the Krebs cycle leading to the accumulation of Figlu in the urine. A 24 hour specimen of urine has to be collected from the patient and this is then tested for figlu using an enzyme which has to be extracted in the laboratory from fresh pigs liver. Charlie spent many hours trying to perfect this technique.

The Department still performed the clinical pathological tests left to it over many years, including reporting centrifuge specimens of urine, working out basal metabolic rates, and examining hot patients stools looking for the first case of amoebic dysentery they had ever seen.

My career was at a crossroads and there were no training posts in Haematology in the country. I had ideas about going to the United States to do research but neither I nor Lumsden knew anyone who could help in this respect. However, I had recently diagnosed three cases of hairy cell leukaemia. This condition had been described first in 1958 by Bertha Bouroncle, Wiseman and Doan. The diagnosis of this condition caused some concern because no one but me in Leeds had ever heard of this condition and I was a Registrar and not a consultant. Professor Tunbridge was under considerable pressure from a bank manager to explain what this diagnosis meant. The matter was therefore referred back to Professor Lumsden who was equally ignorant and so he sent the slides from these three cases off to Professor Dacie at the Hammersmith. He included a letter which stated that Roberts had made these diagnoses and was he right and what did the diagnosis mean? He added that Roberts is at the moment looking for a career in Haematology and would welcome advice.

To my surprise I received a letter from Professor Dacie offering me a job as Registrar in Haematology at the Royal Post Graduate Medical School. This was completely out of the blue and left me in a complete quandary. I would have to find somewhere to live for myself, my wife, and son, in London and at the same time know that there were still no jobs in Haematology in the United Kingdom to move on to. The job was not advertised and was Professor Dacie's gift and was such a tremendous opportunity it could not be refused. I was therefore off on my way to join the best Department of Laboratory Haematology in the world and hope for the best.

Charles Turner Thackrah (1795-1833). A founding father of Leeds Medical School, a pioneer in the field of occupational medicine and one of the first workers in the field of blood coagulation.

Medieval blood letting.

Old Medical School library.

A teaching in the Littlewood Hall in the 1960's. Professor Sir Ronald Tunbridge is standing. Professor C L Oakley FRS is seated to the right. Professor John Cawley, as a student, is second from the left seated on the second row back (see Chapter 8).

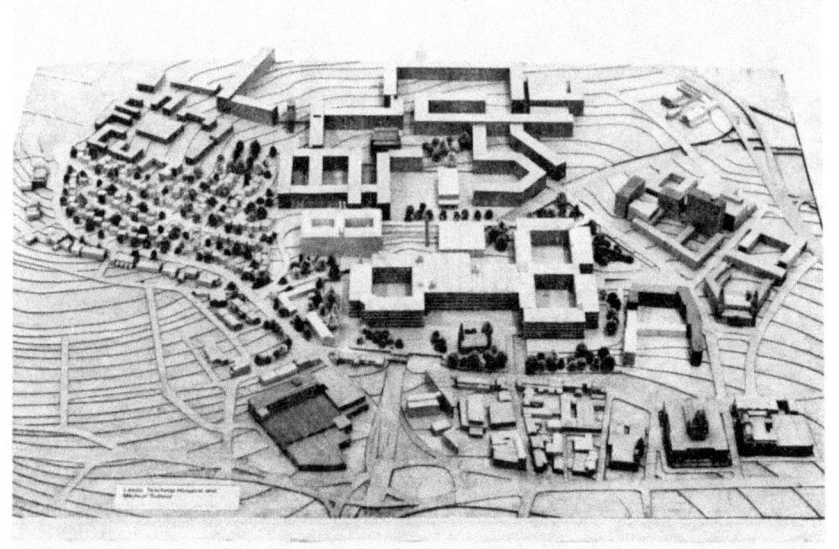

MODEL OF THE PROPOSED NEW TEACHING HOSPITAL AND MEDICAL SCHOOL

A model of the new Infirmary which never came to pass.

The old department of Clinical Pathology which became the Department of Haematology in 1970.

HAEMATOLOGY DEPARTMENT

Work is completed on the provision of a new Blood Bank and the accommodation was occupied on October 6, the department transferring from the Hospital for Women.

The remaining part of the scheme, which provides new offices and laboratories for the Haematology Department, will be finished in late November.

The newly reconstructed Haematology Department (1974) which now accommodates Blood Transfusion (to the right of photograph; the Martin Wing is to the left).

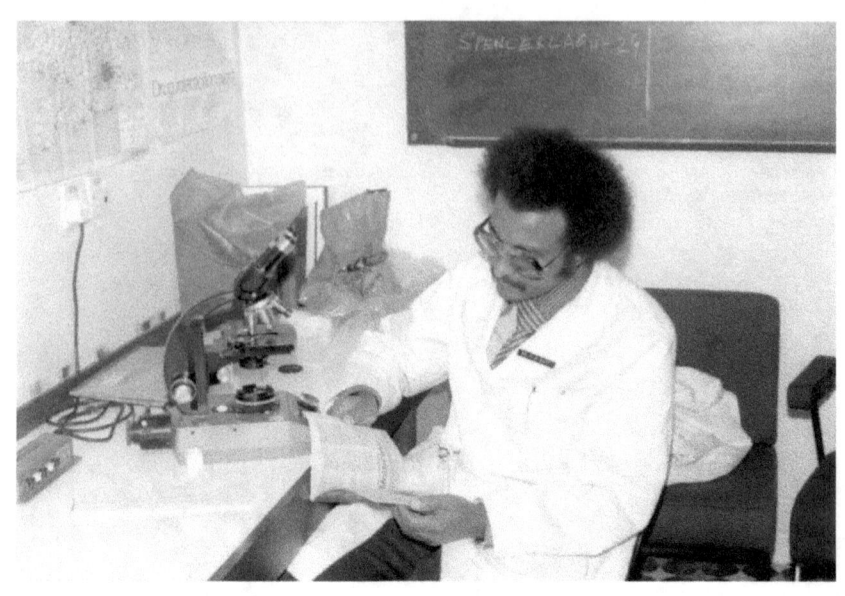

One of our Sudanese trainees.

Mr Charles W Buchan, Senior Chief Medical Laboratory Scientific Officer (1960-1990).

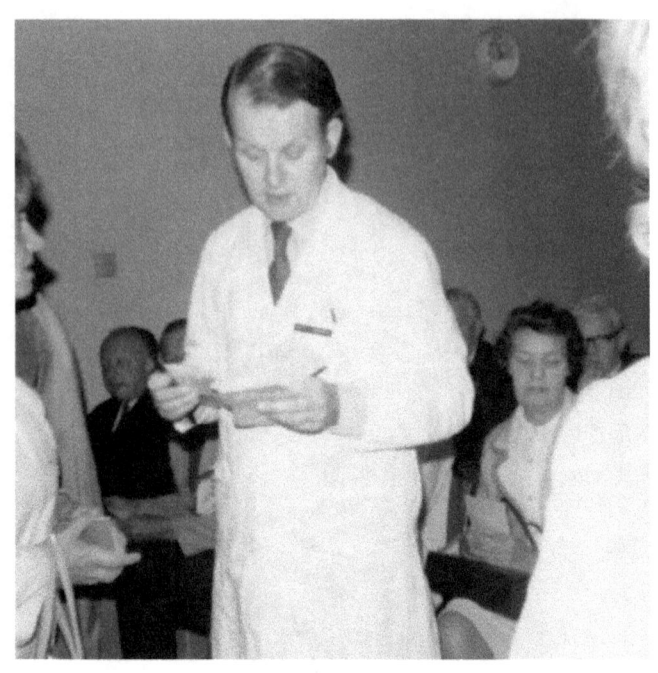

The author at an anticoagulant clinic in the 1960's.

CHAPTER 5

THE ENLIGHTENMENT

The Royal Postgraduate Medical School opened in 1935. It was exactly what its title said, a University Teaching Hospital devoted to postgraduate education and research. It had high ideals which, for the most part, it lived up to. The first thing I noticed as I arrived at the Hammersmith Hospital was that there was a complete absence of Rolls Royce or Bentley cars. I guess most of the cars there had been purchased at second hand.

The core of the Department of Haematology was upstairs off a corridor where there were three rooms, Professor Dacie's office, Room 125, and Room 126 which were Routine Haematology and Special Haematology respectively. There were two NHS Registrars, myself and George Marsh. George was a tall man with a pipe permanently clenched between his teeth and of a strong resolute character. George had been broken hearted when a young woman, the love of his life, left him and ran off with a British racing car driver. To alleviate his heart break he volunteered and joined the expedition to the South Pole, that was being organised by Fuchs and Hillary. In fact George led a team of huskies to the South Pole and planned out the route for Fuchs and Hillary in their mechanised vehicles. (I can imagine George as a Captain of a destroyer leading a convoy through the Artic seas dodging snow storms, ice flows, and enemy U-boats in a journey to Murmansk.) We soon formed a good partnership and worked well together.

Other members of the junior staff were all from overseas on travelling grants and did not have any routine duties. They were from all over the world and I remember particularly Larry Petz from California, Don Cowan from Toronto, Hans Huber from Austria, and

John Lamvik from Norway. George and I had to do all the routine work and participate on the on-call rota. This involved three months rotation through Routine Haematology, Special Haematology, Blood Transfusion, and the laboratory run by David Mollin which was renowned for its work on the assay of B12 and folate. Everyone was very friendly and I was soon at home there. The attitude however was one of continuous challenge, and when one spoke one had to be able to justify everything one said and counter the challenges one faced.

Dacie who was undoubtedly the best Laboratory Haematologist in the world was a somewhat austere and pedantic individual. He was a superb morphologist and diagnostician and his knowledge of haematology was comprehensive but he had a special interest in the haemolytic anaemias. In particular he had a special interest in paroxysmal nocturnal haemoglobinuria (PNH) and, when I was there, had just diagnosed his forty-ninth case. He was essentially a collector; all cases of haemolytic anaemia had a chart with a full collection of results, a blood film, and bone marrow if this had been performed. These were stored in a box file. Samples of plasma and serum from the case were stored in a deep freeze.

Dacie's hobby was collecting moths and he spent most holidays going out at night time with a light, a large net, and bottle of chloroform, indeed he was an accepted authority in this field. I remember one of his Research Scientists telling me that he thought that Dacie had in fact collected him.

In the Routine and Special Haematology laboratories George and I had to perform the routine bone marrow aspiration. All the blood films and bone marrows had to be perfectly spread and stained with a double stain, known as May Grunwald Geimsa, which produced beautiful results. All films had to be spread using a spreader made from a glass slide cut up with a diamond. In fact it seemed that the possession of a diamond was essential working in a laboratory run by Dacie. All films once they had been stained were mounted, labelled, and prepared for examination. On examination of a blood film or bone marrow the film had to be extensively viewed with a dry lens, failure to do this would raise Dacie's ire and, in fact, I worked to this principle for the rest of my life in Haematology. I very well remember some years later applying to the Leukaemia Research Fund for a research microscope. The request was turned down by Dacie, Chairman of the Research Committee, because I had not requested sufficient numbers of dry lenses. All the blood counts in the laboratory were performed according to techniques described in the book *"Practical Haematology"* by Dacie & Lewis but the laboratory had just acquired a new Coulter Model A, a particle counter, which was used to count white cells. Every bone marrow had an iron stain and woe betide any technician who prepared a slide with stain deposits on it. Examining the iron stain of a bone marrow was one of Dacie's keen interests.

All significant abnormalities found in blood films and bone marrows had to be shown to Dacie in the room next door and then at 5.00 pm every day there was a meeting of all staff to look at the problems of the day and it was the job of the Registrar to present these to the staff with comments by Dacie. Dacie could be very critical of any member of staff who did not use haematological terms correctly. On one occasion I remember David Mollin, an international expert in the field of B12 and folate, asking what the retic count was and was told by Dacie to express himself properly "the word was reticulocyte". On another occasion Mollin made some comment about a film and Dacie had turned to Mollin and said "Mollin, if you would sit on the chair properly and examine the film you will see that this is an obvious case of hereditary spherocytosis". Dacie used to ask everybody what they had got to show and Mollin muttered to me one day that he would show Dacie something more than he had expected.

There were other Departmental meetings also. On a Thursday David Galton would run his Leukaemia Clinic and at the end of the day go through all the cases he had seen and demonstrate these on flow charts, and once a month there was a major formal case presentation which was not infrequently attended by distinguished visitors from overseas.

I would sometimes sit down with George Marsh and we would look at slides together and compare notes. George was particularly interested in red cell fragments. Dacie had proposed to him that he should prepare an MD thesis on this topic. Dacie was interested in particular in the relationship of red cell fragments to the Haemolytic Uremic Syndrome and cases of disseminated intravascular coagulation. I was interested in red cells which were round, condensed and spickulated, and called them Sputnik cells as it was about this time that the Russians first put what they called a Sputnik into orbit around the world. Sputnik cells are in fact normal cells which should be removed by the spleen and are a marker, therefore, of post splenectomy blood films.

An onerous task for the Registrar in Routine was to go on the wards and collect specimens for LE cells, that is, cells which occur in the disorder Systemic Lupus Erythematosis. For this, we had to depart to the ward which a conical flask and a rod through a cork at the top of the flask.

The rod was covered in pieces of molten glass(which had been cooled) and this was used to defibrinate blood samples. The defibrinated blood was brought back to the laboratory, incubated at 37^0 and blood films made. These cells were in fact blobs of denatured DNA which had been ingested by neutrophil polymorphs. Search for these was tedious and was often given to the most unpopular member of the Department. In Leeds, we defibrinated blood by shaking paper clips in a universal container.

In Special Haematology we would investigate all the special cases that had either turned up in the Hammersmith Hospital or had been referred from outside hospitals. Larry Petz also worked in Special Haematology, he had a special interest in Serology and Dacie had him investigate the antibody characteristics such as specificity and thermal amplitude in several cases of autoimmune haemolytic anaemia he had in store. Dacie would check every step he had performed and on many occasions Larry had to stop and start all over

70

again because he had not got the temperatures correct. Larry eventually went back to the United States to run Blood Transfusion in California and took George Garitty from Blood Transfusion with him. (Over the past year I noticed that in the Journal of the Royal College of Pathologists that both Larry Petts and George Garitty had been given life time awards for services to Blood Transfusion in California.) Dacie himself was an outstanding technician and could pull perfect Pasteur pipettes. What he also took a delight in doing was to make automatic delivery pipettes from glass tubing, a thin capillary tube pulled on making a Pasteur pipette, and a stick of ceiling wax.

Sheila Worrledge who ran Blood Transfusion, once told me that Dacie appeared in the corridor outside his room in a very agitated state. His appearance was that of a footballer who had just scored the winning goal in the last minute of extra time in the World Cup. He was shouting "look, look, look" and lots of worried members of staff all went into his room. There was a glass tile with some agglutinated red cells on it, in fact, this was the first positive Direct Antiglobulin Test he had performed with reagents made by himself by immunising rabbits with immunoglobulin! In fact, the Department of Haematology continued to make many of its own serological reagents.

One day in Special Haematology, Dacie took me to see a patient who had been referred to him with a possible haemolytic anaemia. The patient was the wife of a foreign diplomat and was probably of middle eastern origin. I was presented with a large glass syringe and a large gauge metal needle for me to perform a venesection. I had to let the venous pressure of the patient push the plunger back so that no bubbles ensued. In actual fact one bubble did appear to a loud "tut, tut, tut's" from Dacie who pointed out that if this bubble burst it would rupture red cells, release haemoglobin and raise the plasma haemoglobin level. At the completion of the venesection I handed the syringe to Dacie who then proceeded to fill several tubes. Some were allowed to clot to provide serum and others plasma, and a blood count was performed. I had to separate and collect the plasma and serum and send it for the special tests required. I then froze the remainder of the sample. Some days later Dacie asked that I get a further sample of serum from the deep freeze for this patient and I told him that we had run out. I then got a dressing down from Dacie who told me that one never, ever ran out of a specimen. If there was only 1 ml left then the test had to be performed on 0.5 ml and so on. A specimen was sent for routine tests to Chemical Pathology and it was well known at the time that Chemical Pathology technicians should they spill the specimen, break the glass tube, or feel generally out-of-sorts, would write "inad" on the result form, meaning inadequate specimen. Dacie filled the Chemical Pathology tube himself and wrote the label. Thus the tube had been filled and labelled by Professor Dacie one of the most renowned Laboratory Haematologists in the world who would be, in years to come, a Knight of the Realm, Fellow of the Royal Society, and President of the Royal College of Pathologists. When he received a report from the Department of Chemical Pathology with "inad" written on the bottom his face went puce. He became extremely agitated, went into orbit, and descended upon the Department of Chemical Pathology. There were no reports of any fatalities but it must have been a close run thing.

On another occasion a case of putative PNH was referred to Dacie for diagnosis and management. I was deputed to perform the acid lysis test, or HAM's test, with Dacie standing over me. Pipettes had been left in a somewhat disorganised state on the bench and I picked up the wrong pipette and dispensed the incorrect amount of hydrochloric acid into the mixture and the red cells went black. An agitated Dacie then hit me on the head several times with a pipette stating "you are no longer in Leeds now, Roberts". I resented this and swore that I would get my own back which eventually I did.

Years later, in Leeds, a bright young doctor by the name of Peter Hillmen joined us from the Professorial Medical Unit. Pete was obviously very able and bright and was advised to get some experience elsewhere and joined Lucio Luzzato at the Hammersmith where he had replaced Professor Dacie. Pete was given a project by

Lucio Luzzato, which Lucio confessed to me later he did not think would work, but Pete made it work and discovered the basic defect in paroxysmal nocturnal haemoglobinuria, a red cell skeletal defect. His work was published as a plenary paper in the New England Journal of Medicine and his co-authors were Luzzato, Lewis, and Dacie, a very distinguished panel indeed.

This was a major achievement and I was determined to get Pete back to Leeds but I had to wait for a Senior Registrar vacancy. Whilst working with us he successfully passed the final MRCPath and my next job was to find him a consultant post. Mike Galvin, Consultant Haematologist, at Wakefield had some sessions and was keen to have a joint appointment. Derek Norfolk had been appointed Research Director at the Infirmary which gave me three sessions to join up to those at Wakefield. However, Stuart Ingham, the Chief Executive, did not agree and I had a terrible fight to squeeze out these three sessions from Stuart Ingham, but finally did so, and we were very pleased to appointment Peter Hillmen as a Consultant.

Ultimately, Peter moved full time to Leeds and it was then that he had the idea of testing the effect of a monoclonal antibody against complement in PNH. The treatment worked and this represented another major triumph for Pete. The antibody, however, proved to be very expensive and meant that it should be administered nationally. Leeds was then appointed a National Centre for Diagnosis and Treatment for PNH.

Every time I hear that Leeds is a National Centre for the Diagnosis and Treatment of PNH a smile appears on my face and until now nobody ever knew why.

The key to much of Dacie's work was the collection and storage of specimens with related completed investigations. The value of this was seen when a paper appeared in the literature reporting that haptoglobins were absent from the blood in five cases of haemolytic anaemia. Dacie was able to repeat this work in a large number of cases and relate the haptoglobin level to the red cell survival. Dacie therefore wrote the definitive paper.

In the late 1990's the then Labour government fanned the flames of hysteria over the retention of tissue samples. Some mothers of children who had died from congenital heart disease complained that their children's hearts were being retained without permission. This complaint was then extended to include any sample of tissue stored in the laboratory. The atmosphere was then of a medieval witch hunt totally out of context with our modern scientific society. In practicing pathology we were expected to keep specimens from interesting cases for use in teaching and no special permission was obtained. In the ensuing maelstrom many specimens had to be returned to relatives. I had already retired and observed all this with dismay thanking God that I was out of it.

I do know that if the legislation of the Human Tissues Act had been retrospective I would have spent several years in jail but I would have had the company of Dacie and many other illustrious colleagues.

My rotation to Blood Transfusion was very interesting. I often helped George Garitty in the bleeding of rabbits and participated in the investigation of antibodies. I learnt the basic techniques in transfusion which I needed to perform my on-call duties. Hammersmith Hospital is a tertiary referral hospital and did not take in many acute cases and the demand on the on-call service were few.

A further three months was spent with Mollin in his laboratory which specialised in the assay of Vitamin B12 for which his assistant, Barbara Anderson had a formidable reputation, and Folate. The Folate assay had recently been developed in this laboratory by Dr Alan Waters, an Australian Research Fellow, who I would get to know well. Alan had found that heating serum to precipitate protein destroyed folic acid making the assay impossible. Alan discovered that the addition of ascorbic acid before heat treatment of the serum preserved the folate and thus a folate assay was possible. This assay was an enormous advance in the investigation of nutritional anaemia and performed all over the world. It further enhanced the reputation of this laboratory.

Alan was to begin to write many papers on serum folate in various disorders. When he left Victor Hoffbrand replaced him and continued to write innumerable papers on the significance of serum folate levels in many diseases. Mollin was an interesting character; one could turn up in the laboratory one day and find him strolling around with a large cigar in his mouth looking like a Texan oil magnate. On another occasion he would be found on the floor of his office, prostrate, his head on a cushion, suffering from migraine. He was extremely temperamental and on one occasion, before I did my rotation to join his Department, Michell Lewis sent me with some slides I had to photograph to Mollin asking for permission to use his microscope. Mollin went berserk because he had a very modern microscope with a large control panel that he had not the slightest idea how to work. I did however get into Mollin's very good books when I looked at a film, found Sputnik cells, and told Mollin that this patient had an absent spleen. He was highly delighted because this was from one of Professor Chris Booth's gastroenterology patients. Mollin worked closely with Chris Booth and would be able to tell him that the haematolgists had made the diagnosis of adult coeliac disease for him.

On another occasion a megaloblastic anaemia was admitted onto one of the medical wards. This was a rare event about which Mollin got quite excited. The patient, to begin with, had a bone marrow, B12 and folate assays, and was then put on a low dose of B12, and daily reticulocyte counts for ten days. The patient then had a further bone marrow, further B12 and Folate assays, followed by a very low dose of folic acid daily, and the process repeated. At the end of 20 days a B12 absorption test was performed according to the method of Schilling which involved giving the patient a large dose of B12. The study was to show that with present conventional doses of B12 or Folate, a patient deficient in one may respond to the other but not if sufficiently low doses are given. Thus the patient, who would not be admitted to hospital at all in this day and age and who would be treated by a General Practitioner, had over three weeks in hospital. However in some ways this was ground breaking research.

I also spent some time with Michell Lewis in the Isotope Department. I learnt to do a blood volume but as there were no suitable cases that needed red cell survival I learnt nothing of this technique.

The person at the Hammersmith Hospital who had the most influence on me was David Galton. He combined his keen clinical interest in that of an assiduous investigator of the pathogenesis of malignant haematological disease. He was one of, if not the best, diagnosticians of leukaemia in the world. To work with him was a pleasure, he was shy and self-effacing at times but intellectually incisive. In my view this combination of clinical interest and laboratory practice is what the discipline of haematology is all about and he embodied this.

David Galton was the son of a Hungarian general practitioner who emigrated to England. David was born in London and was educated at Cambridge University and the University College Hospital in London. After his house jobs he got an appointment at the Royal Marsden Hospital and Chester Beatty Institute under the direction of Alexander Haddow who oversaw a program of the synthesis and testing of anti cancer drugs. David Galton was given the task of evaluating the new drug busulphan in chronic myeloid leukaemia and chlorambucil in chronic lymphocytic leukaemia. When I first knew him, he was a Honorary Consultant Haematologist at the Hammersmith looking after patients with leukaemia and lymphoma.

Galton and Dacie were good friends with several mutual interests. Both loved classical music and it was said that David Galton forsook a career as a concert pianist to pursue his career in haematology. Dacie played the piano every morning before coming to work. Dacie was a distinguished lepidopterist and Galton was a passionate ornithologist. I remember at a conference in Cambridge seeing David Galton looking up various nooks and crannies between buildings trying to find house martins nests that he knew of when he was a student there. He finally retired to the Norfolk coast to pursue his love of birds.

Dacie was the first editor of the British Journal of Haematology and it was said that he, by virtually re-writing every paper that was sent to the Journal, established it as one of the best haematology journals in the world. David Galton was the second editor. I remember submitting a paper on some work which was to be part of my MD thesis for publication. David Galton's constructive criticism of the work was a very great help in the preparation of my thesis.

I kept up my acquaintance with David Galton through the Medical Research Council Working Parties on adult leukaemia and through the British Society for Haematology meetings. When the Medical Research Council created a Leukaemia Research Unit at the Hammersmith he was appointed as its first director. He influenced many bright young people who had chosen a career in the field of leukaemia.

My career in haematology was very much influenced by him and that influence has persisted in the Department of Haematology until the present day.

Life in general at the Hammersmith was very interesting. There were two major open general events. One was the fact that each lunch time a post mortem demonstration was performed. The case would have a post mortem performed by a Registrar in Pathology who had to present it but this would be preceded by a case presentation from the Houseman looking after the patient. It should be noted that at the Hammersmith a House Physician was of Registrar status anywhere else. The procedure was observed by the Professor Pathology, Professor Harrison, and the Professor of Medicine, Sir John McMichael, who made comments. The radiology from the case was presented by the Professor of Radiology Professor Steiner. The post mortem room was packed every day.

The big event of the week however was the staff round held in a large theatre attended by many Professors of Medicine from London. Again the cases were presented by the Houseman from the Firm with comments from the boss. Cases were debated and should anybody make an unsubstantiated statement they would be cut to ribbons. I very well remember meeting a former colleague, junior to me, from the Infirmary walking along the corridor one Wednesday morning. He told me he was training to be a Cardiologist and Geoffrey Wooler had written to the Professor of Cardiology, Professor Goodwin, recommending him for the post of Registrar. The Professor had written to this young man suggesting that before the interview he attended the staff round to know what life was like at the Hammersmith. He had therefore come the night before, got to the Hammersmith in time to go to the staff round. The Registrar presented this cardiological case and there were then comments from the Professor Goodwin who pointed out that this was an interesting case of mitral stenosis for which Wooler, in Leeds, had a special operation. He then said that the meeting was fortunate in having a member of Mr Wooler's team present and would the young doctor please come onto the stage and tell the audience about this interesting condition. My colleague from Leeds then proceeded to be interviewed by the leading Professors of London but did well and got the job.

Life was interesting at the Hammersmith and there were some unusual characters in that the Consultant Neurologist was an anarchist, and the Chief Obstetrician was a Physician. I found virtually everyone very friendly and one day a Senior Physician came to speak to me about a case in Routine Haematology and we discussed it on friendly terms and one of equality. The visitor was in fact Professor E L Bywaters, a leading world authority in the field of rheumatoid arthritis. On another occasion I spoke with Mr Wallet, Senior Chief Technician in the Department of Haematology, and said that I would like to go to the international match at Wembley between England and Brazil. In two or three days he produced a ticket and said that he had fixed me up with a lift and that I should turn up at a specific corner of a building to meet the other members of the team, in fact, these turned out to be two Professors, and a Reader. There was no problem of seniority whatsoever and I was treated as an equal; this would have been impossible in Leeds.

However, time was marching on and I had no job to go to at the end of my year at the Hammersmith. As far as I could see there were only two places where I could continue a career in Laboratory Haematology. One was the University College at Lagos where Sheila Worrledge had spent some time, and the other one was the University College of the West Indies where John McIver had worked. I was about to write to them to enquire about vacancies when I got a letter from Professor Lumsden inviting me to apply for a post of University Lecturer in the Department. What had happened was that McIver had left and taken a post of Consultant Haematologist at Manchester. This was the vacancy I would fill. The story I heard from McIver later was that he had been to the Committee of Physicians (there was no Committee structure representing Pathology in the Infirmary) and had negotiated for the screening of urine and stools for parasites and cysts to be taken over by the University Department of Microbiology. Lumsden was apparently incensed and tore into McIver for doing this without his permission. This however was obviously to the benefit of the Department of Haematology but McIver's position was now untenable. Thus I was homeward bound to fill a post that I thought would be a Lecturer in Haematology but I was to be disappointed.

I had however first experienced the most formative and exciting year of my medical career. The breadth of experience in haematology in Leeds was actually greater but haematology in the Hammersmith was stimulating and exciting and performed to a very high standard. I had also worked with some of the best haematologists in the world. My challenge now was to reproduce this in Leeds.

CHAPTER 6

A WAR OF ATTRITION

I arrived in Leeds thinking that I was to be a Lecturer in Haematology. When I arrived in the Department of Pathology, I found out that I was on the usual quarterly rota of one month post mortems, a fortnight or more as it turned out in Histology, and the rest would be my own time, but obviously time that I would spend working in Haematology. I went over to the Department and met the technicians who were as pleased to see me as I was pleased to see them, and two new Registrars who had been appointed in my absence. One was Cedric Abbott, the other Colin Merry, and both told me that Professor Lumsden had made it quite clear that I had no seniority or authority over them whatsoever. It was quite obvious that Professor Lumsden saw me as a threat and was determined to keep me under control. In this he was absolutely correct. I was determined to promote myself within the Infirmary with a view that at some stage I would be a Consultant. The battle lines were therefore joined, but Professor Lumsden had struck the first blow.

It was announced about this time that there was to be a College of Pathologists and that all those who were interested in being a member should apply. The way that it worked was that all existing Consultants would be given membership on the payment of a suitable fee, while those in training had to apply. The criteria for those in training would be determined by the Heads of Departments they were working in. Most of my contemporaries had to take the final exam only but, no doubt on Professor Lumsden's recommendation, I had to take the primary exam which was designed for someone who had about two years' experience. This was no joke as to fail this exam would have been very ignominious for me and the other thing was there were no such things as travelling expenses or examination expenses, all this would have to come out of my own pocket. I proceeded down to London to do the exam. Practical examinations were conducted in large laboratories designed no doubt for examining sixth formers. The haematology was not difficult but it was easily possible to drop a tube or make some mistake to fail. The histopathology was easy. I managed to get through this exam with no problem and so proceeded within a few weeks to take the Final examination. The written papers were held for people living in the North of England in Manchester. Feeling very sorry for myself I booked in at the Midland Hotel in Manchester which was the best hotel in that city and enjoyed dinner listening to a stringed orchestral. The next morning as I awoke, the room span round, I felt dreadful, and obviously had contracted a severe food poisoning. Notwithstanding I proceeded to the written examinations which I got through.

The practical was held in Sheffield in a Department run by Professor Eddie Blackburn, a Leeds graduate, and as I got to know him later a very nice person indeed. I found the examination very difficult in that most of the blood films were swimming in oil, reticulocyte preparations and iron preparations were dirty, and since I had been working at the Hammersmith recently found this appalling. There was no question of using the high dry lenses on the microscope as these were immediately covered in oil. The Blood Transfusion examination I found difficult. I was told by many that in an examination one should never do blood groups on tiles, however with this examination unless one actually did take short cuts by doing blood groups on tiles it was impossible to finish and so I failed it.

I then proceeded to take the repeat examination in the Autumn. Much to my surprise the secretary of the Department shouted to me, "there's a phone call for you here, its from the Registrar of the Royal College of Pathologists". Somewhat surprised I went to speak to him and he asked me if I would go down to London for the written as they had devised a multi-choice examination and because it was new would have to be supervised. I went off to London and my good friend George Marsh from the Hammersmith put me up for the night.

After a pleasant dinner with George and his wife, George suggested that I go for a walk with him on Hampstead Heath. We then proceeded outside and George poured his heart out to me and told me that he had fallen in love with Mollin's secretary at the Hammersmith and was determined to leave his present wife whom he found very difficult. George needed to speak to someone out of town who could not pass any of these stories on and no doubt found it a great help to speak to me, but the end result was that I did not get to bed until 2.00 am. I was awoke next morning by a low flying jet at 6.00 am so I had a limited night's sleep but proceeded on to do the multi-choice examination which I found very searching but an excellent exam.

I got through the multi-choice and went to do the practical exam in Glasgow. The two examiners were Dr Hutchinson from Glasgow, and Dr Nelson from Belfast. The blood transfusion examination on this occasion was very fair. We had to identify an antibody but also found an unexpected cold antibody of which the examiners knew nothing. Having got rid of that and being much relieved I then proceeded to the coagulation examination which, as I have pointed out, consisted largely of a thromboplastin generation time. I proceeded forth with confidence but much to my incredible horror found out that nothing would clot. I stood up in frank disbelief, very agitated, and looked round and found that my co-examinees were also in an equal state of distress. Dr Nelson came along and looked and did not quite believe us but nevertheless we sat down and all the material was removed and we were presented some time later with a new test. We proceeded forth once more but to no avail, nothing again would clot. The examiners were somewhat perturbed about this and I told them that what they had done for this examination is use a serum which had been frozen in the deep freeze and immediately thawed it rather than leaving it overnight to activate which was essential. They did not really want to know but fortunately for us the young woman in the examination was from Belfast and was known well by Dr Nelson and had completed an MD on haemophilia in Northern Ireland and had done many, many thromboplastin generation tests and factor assays, and if she could not do the exam then nobody could. We were therefore told that we had demonstrated our skill in coagulation practical and that we were through this part of the examination.

Next day the morphology slides were easy and so we all went home in the knowledge that we probably passed the examination which, in fact, we had. One of the candidates at the examination was Colin Geary from Manchester whom I subsequently got to know well and would play a further part in my career which you will find out at the end of this chapter.

I therefore went back to work a much happier person knowing that I had a ticket which I could use to get out of the clutches of Professor Lumsden any time I wished to use it. I was very unhappy in my present post, the relationship between a member of staff and the Head of Department was feudal. A Professor was absolute boss and there was no recourse to any other body should there be a dispute of any kind. Keith Anderson, the Senior Lecturer in Urological Pathology, had a dispute with Professor Lumsden and wrote to the University. In fact for this purpose the University did not exist. His letter was passed immediately to Professor Lumsden without anyone else ever looking at it. The power of the Professor was absolute and he could move anyone out of their room, take away their equipment, and make their life generally impossible, but there was no recourse to justice and no one to resolve any dispute. I remember a Pathologist with a personal chair in Manchester taking three months' compassionate leave to look after his dying wife. When he got back to the Department his room had been completely cleared out, the contents were dropped in the middle of the corridor, and somebody else was using his room. There was absolutely nothing he could do about it. Very often when someone wanted to leave, if that person was useful to the Professor then they would get a poor reference. It was therefore a form of tied employment and thoroughly unacceptable. Therefore by having the membership of the College of Pathologists I had the means of leaving Professor Lumsden and his Department should a suitable job turn up.

Professor Lumsden however hardly ever visited the Haematology Laboratory. He managed it by regularly seeing Charlie who would speak to him every now and then particularly when he wanted some more money to buy an item of equipment. I was therefore, with the help of Charlie, able to begin to make changes to try and bring the laboratory up to Hammersmith standard. I introduced the use of spreaders to make films, got the technicians to write labels, and store slides after they had been used and to stain them with May Grunwald Geimsa stain. The slides therefore were beginning to look like those at the Hammersmith but in Leeds the volumes of material were greatly in excess of those at the Hammersmith and there would be simply no time to mount the slides with a cover slip and wait until they had dried to report and store them.

A process of evolution was taking place with regard to syringes and tubes. A central sterile Supply Department was set up in the Infirmary and began to sterilise syringes which were then available in large packs. As time went by we entered the plastic revolution and we would subsequently have disposable syringes and needles. This meant that syringes and needles were now supplied by the Hospital and were no longer a Haematology Laboratory responsibility.

With the appointment of more and more phlebotomists it soon became rarely necessary for a technician to visit the wards unless it was for a special sample, such as, blood for coagulation studies. Specimens were sent in to the laboratory and a Specimen Collection Porter was appointed. The gentleman concerned was a Mr Charlie Biles who had escaped from Latvia after the war and moved to England. His base was the Rest Room of the morticians just round the corner. Some of the morticians concerned were Arthur who had recently been appointed, and Victor again an ex patriot Latvian who had come to this country and been appointed a mortician. He had the misfortune to have a severe motorbike accident which fractured his skull and he lost his sense of smell. This was however of great benefit to him as a mortician but he notwithstanding obtained a substantial sum of compensation.

I got to know Charlie Biles very well and he had a remarkable history. In the 1936 Olympics in Berlin, often called the "Hitler

Olympics", Charlie represented Latvia as a weight lifter and successfully obtained a bronze medal. He was a hero in Latvia when he returned but he realised at the outbreak of war that he would be a target for the Russians when they invaded Latvia and so he made his way to Germany with his wife where they were separated as Charlie was recruited in the Wehrmacht, the Germany army. Charlie fought on the Russian front and on one occasion crept up to a Russian machine gun post and wiped it out with a hand grenade. For this he was awarded the Iron Cross by Hitler. Charlie was a very avuncular pipe smoking man and had been fortunate enough to be reunited with his wife as they entered England. His wife had a job in pharmacy. His life however was cut tragically short when he was in his 50's when he had a sudden myocardial infarction at home.

Victor and Arthur were again interesting characters. Victor, despite his loss of smell, hardly seemed to appear in the post mortem room. He donned a white coat and interviewed patients coming to collect a death certificate. He sat very seriously behind an imposing desk in an immaculate clean white coat and turned over the pages of the notes telling the patient's relatives that this case was very important to medical science and certainly needed a post mortem. If the patient's relatives refused then Victor had no hesitation but to refer the case to the Coroner. Thus the post mortem rate at the General Infirmary at Leeds was very high indeed and we were all kept very busy in the post mortem room. I would very often do five post mortems in one morning just to keep up with the pressure of work. The point of this however from Victor's point of view is that the morticians were given a bonus payment for every post mortem they did, so every post mortem he managed to get permission for represented a further increase in his salary. Another perk was the fact that Pathologists were asked to remove the pituitary gland and drop it into a large bottle of formalin. For this the morticians were paid 50p. The tragedy however of the collection of pituitaries in this way was that the growth factor which was extracted from them was contaminated with neurotopic viruses leading to unfortunate infections in young patients, so much so that the practice was abandoned.

Arthur was an impressionable character and I remember when he started, Keith Anderson, the Senior Lecturer in Urological Pathology, had a word with him and told him how he had to behave. Arthur therefore turned up at the front door of the Infirmary in an immaculate overcoat with a velvet collar, a rolled umbrella, and shiny black shoes. He was told to approach the lift at the side of the steps leading up to the bust of Moynihan, twirl his umbrella and press the button of the lift with it. Arthur however balked at the final request that was he should wear a brown bowler hat. He also came into the rest room and ate German sausages just like Victor did. Arthur had a fine baritone voice and while we were all working in the post mortem would regale us with tunes from the operas, dispersed with the odd mucky tale that he had just heard from the undertakers. There were many tales about Arthur, but I was most surprised some years later when in the middle of a program of "Songs of Praise" from Leeds he came onto the programme as a mortician who had found God; how long this lasted I am not altogether sure.

Conditions in the rest room where the technicians and Charlie Biles had their lunch were appalling as it immediately abutted onto the gents toilet. I now understand why Professor Stewart when helping to design the Algernon Firth Institute organised personal en-suite facilities.

I thought, bearing in mind my experience with Mollin at the Hammersmith, that I should introduce microbiological assays for B12 and Folate. For the B12 assay one needed a small aquarium tank and a thermostat which would cost very little. Charlie went over to see Lumsden and asked him for the money to buy this equipment and was told by Lumsden that he did not given money to Boy Scout pathologists.

It was becoming apparent to the Department that there was an increase in the number of one stage prothrombin times being performed on patients on anticoagulant therapy. At this time we made the thromboplastin ourselves in the Laboratory. A one stage prothrombin time consists of a sample of plasma being added to some material called thromboplastin and the time of coagulation

86

measured after the addition of calcium. It is in essence a very simple test but the problem is the thromboplastin. The way we made thromboplastin was to get a brain from the post mortem room, strip it of blood vessels and membranes and then mash it up with a machine that I had bought to mash up the food for my son shortly after weaning. The brain was mashed and dropped into acetone, stirred up, and then spread out to dry on a large table covered in blotting paper. The dried material was then stored and an aliquot removed and incubated with saline before use. The normal prothrombin time is about 12 seconds but with each batch of thromboplastin there is a considerable variation in this. Patients on anticoagulant therapy with dicoumarol type drugs is prolonged and has to be controlled by varying the dose of the drug according to the prothrombin time. A variation in prothrombin time due to the thromboplastin therefore makes management of the anticoagulated patient difficult. There was a commercial thromboplastin available called a Thrombotest and this was demonstrated to Professor Lumsden but he refused to fund it.

There was one alternative available, a test called the P&P test, a Prothrombin and a Proconvertin test, in which two clotting factors are added to the thromboplastin to make it less sensitive, or sensitive mainly to the factors concerned in dicoumarol therapy. For this, Charlie and I had to visit the abattoir in the centre of Leeds to obtain some ox plasma. The abattoir was situated just behind the bus station in Leeds and we entered the abattoir to find many cows in various stages of decomposition. A cow then entered the door to have a quick look at all its former friends before a humane killer was applied to its head and it dropped immediately unconscious on the floor. Straight away chains were applied to the back legs, the cow was hoisted rapidly upside down and its throat cut. Charlie then had to step forward with a Winchester bottle with anticoagulant in and a funnel to collect the ox blood. As soon as we had finished the black pudding lad stepped forward with a large tray and filled that. This was one of the most shattering experiences of my medical career, far worse than the first time I went into the post mortem room. I was so shaken that it made me buy a packet of cigarettes and start smoking again. We used the test and it appeared to work but the thought of making regular visits to the abattoir to get further reagents put us completely off. The problem therefore of considerable variation between batches of thromboplastin continued. If this was a problem for us in the laboratory it presented an even greater problem for medical staff adjusting dosages of anticoagulant therapy on the wards and in the Out Patient Department. A decision was made therefore to commence Anticoagulant Clinics which would be run by the Laboratory.

Charlie and I therefore departed to do a clinic run from the phlebotomy suite. Charlie would go with a trolley, water bath, and reagents, and set up a mini laboratory, whilst I took the results and called the patients. We devised a record system of small cards in a plastic wallet which recorded date, prothrombin time, and dosage, and a similar series of cards for laboratory records. Before the clinic began the clinic list was obtained and the record cards removed from our store. Eventually a part time voluntary lady came and did this job for us. The clinics progressed efficiently and I expressed no worry about coping with the numbers because at the time most of the patients were being treated for myocardial infarction and the anticoagulant therapy was to prevent recurrence. Recent publication of statistics however showed that the anticoagulant treatment for this particular condition was not very satisfactory, in fact probably a waste of time. I expected therefore the clinics to progressively diminish in number: how very wrong I was. To cope with the variations in thromboplastin activity I would collect together the first ten results and compare them with previous results and one could see from the review of the results on patients who had a very stable record what the performance of the thromboplastin was and I could make various mental adjustments along the way in adjusting dosage.

Again, this was hardly satisfactory but a white knight of the anticoagulant service appeared in the form of Leon Poller, Consultant Haematologist at the Withington Hospital in Manchester. He ran the National External Quality Assurance Scheme (NEQAS) for anticoagulant therapy. At first he mainly obtained lots of results to assess performance of locally produced thromboplastin. There was no doubt the problem of inconsistency was widespread. Not only were there local quality problems but international ones too. In particular the performance of locally manufactured human brain thromboplastins compared to commercial animal brain thromboplastins was markedly different leading to very different levels of anticoagulation of patients which could lead to haemorrhagic problems. The International Committee for Standards in Haematology took an active interest in this and a method of calibrating thromboplastins was devised. Poller was then making significant amounts of human brain thromboplastin in Manchester. He calibrated this according to International Standards and sent samples round to all NEQAS participants to calibrate their own thromboplastins. As time went by Poller further improved the situation by making enough human brain thromboplastin, which was calibrated according to International Standards, to supply most hospitals in England. As this was performed in National Health Service laboratories on local human material there was no charge. This was obviously a boon for all users and lead to a great improvement in the quality of anticoagulant therapy. A prothrombin time was now expressed as the ratio of the test result over the control result which was called the International Normalised Ratio. This went well for some years until the Aids epidemic and the problem of other neurotopic viruses. The manufacture of human brain thromboplastin had to stop and Poller reverted to a rabbit brain thromboplastin and very little difference was noticed by local hospitals except that there was now a charge for this material as it was animal and not human.

In September 1964 a vacancy for a technician appeared, and following the advertisement we appointed Geoffrey Tate who had been born in the same village as my mother and went to the same school as I did. Geoffrey was the first male technician to be appointed for many years. When Charlie went to see Professor Lumsden for confirmation of the appointment Professor Lumsden turned awkward and said that he did not believe that there should be male technicians in Haematology, all technicians should be pretty and attractive to cheer the patients up, not knowing that in these days most of the bloods were collected by phlebotomy technicians.

Charlie rebelled and threatened to resign if he did not get his own way over this appointment which eventually he did. Geoffrey would become a future stalwart of the Department and took a keen interest in coagulation.

Meanwhile, other developments were taking place in the Laboratory. We purchased some equipment which would suck up a sample of blood and blow it out with a fixed volume of diluent which eliminated all the mouth pipetting that was taking place. A major development in Haematology was the introduction of automatic cell counters. There was a parallel development of automation in the Department of Chemical Pathology. One development depended on the discovery of bubbles. The Technicon machine sucked a column of diluted blood along a thin capillary tube and different samples were separated by a bubble. This was ideal for Chemical Pathology but was also useful for automated haemoglobin readings. In Haematology we decided to purchase a Coulter Counter. This embraced the principle of particle counting which involved sucking up a suspension of cells through a tiny orifice. An electric current was flowing through the diluent and through the orifice and every time a particle went through the tiny orifice the electric current was disturbed and these disturbances were then counted.

When I was at the Hammersmith they had a Coulter Model A but in approximately 1965 Charlie persuaded Professor Lumsden to buy a Model D which can be seen in the illustrations [page127]. Thus we were able to make automatic dilutions of blood for the Counter. The samples then were presented manually to the Coulter D which then gave us a particle count. This really therefore was the beginning of mechanisation rather than automation of Haematology but enabled us to cope with the ever increasing burden of large numbers of blood counts.

About this time Cedric Abbott, a Registrar in Haematology, was approached and asked if he would provide a Chromosome Diagnostic Service for the paediatricians and obstetricians. Cedric sat down and after a short period of time began to produce results. This was not an easy technique to set up. It involved treating a sample of blood with red bean extract to stimulate the cells into

90

division, arresting the cells in the metaphase stage of mitosis with colchicine and then staining them, photographing them, and after magnification cutting them up with a pair of scissors to divide them up into their constituent pairs, and identify any abnormalities. Cedric soon identified a benign abnormality, the Christchurch Chromosome, and things appeared to be going well when Professor Lumsden appeared in the Laboratory and told Cedric to stop as a Regional Service was to be provided from St James's, not that St James's had a Chromosome Service running at the time.

Cedric appeared remarkably undisturbed by this major disappointment and was soon approached by the Urological surgeons who asked Cedric to set up a Tissue Typing Service as they wished to commence renal transplantation. Cedric then, with the permission of Professor Lumsden, departed to Cambridge with Leslie, one of the technicians, to learn the technique of tissue typing. On his return from Leeds and with the help of Leslie, a very skilled technician, they soon had the system going only to be told a few weeks later by Professor Lumsden that this service was to be provided on a regional basis by the Regional Blood Transfusion Service. These were major blows and it was indicated that Haematology had a bleak future with no significant developments. Cedric took these major blows with equanimity - I do not know how.

I was still pursuing the same course as Acting Consultant Haematologist to the Infirmary as well as performing post mortems for one month, and Surgical Histopathology for two weeks in each three month period. This provided little opportunity for research. A month performing post mortems was very busy. I would often perform as many as five post mortems in one morning and spend most of the afternoon writing them up. One factor however did help a little and that was I performed reports for the Coroner and occasionally attended Inquests and this meant I received some fees. The problem was I got absolutely no job satisfaction from performing these post mortems and from time to time these could be a little upsetting. I particularly hated having to do post mortems on children and babies. To pick up a dead baby I found disturbing.

Some of the Inquests could also be difficult to handle. I well remember performing a post mortem on a young woman who had apparently died suddenly. When I opened the abdomen she had peritonitis and in fact died from piercing her uterus with a knitting needle in an attempt to perform an abortion. The Coroner's Court was full of distressed relatives who had no idea that their daughter was pregnant and, as far as they were concerned, did not even have a boy friend. On another occasion a young girl had swallowed her mother's tablets and died, this not only upset her parents but drew the attention of the press and I had several enquiries from newspapers asking for details of the case. The Coroner's Court ruled however that no crime had been committed. Perhaps the alleviating factor about my relationship with the Coroner was when I was invited to the annual Christmas lunch which consisted of abundant booze, large numbers of pork pies, and ham sandwiches; I then mixed with hard bitten gnarled detectives from the local CID and crime reporters from the local newspapers.

In the midst of my unhappiness and despair came a shaft of light. I was asked to meet with a Sub Committee of the Priorities Committee of the Board of Governors. This was a Committee which met annually to allocate capital to various projects. I met with Dr Stephen Anning, Secretary of Faculty (the Medical Advisory Committee) and notable Infirmary historian, and Dr Derek Taverner, Reader in Medicine. They told me that the Infirmary was proposing to set up a National Health Service Department of Haematology and would I hang around and apply for the Consultant post. I was overjoyed as this would realise my major ambition.

However following this when I heard no more I went to see Arnold Tunstall, the Hospital Secretary, who told me that he had written accordingly to the University to inform them of the Infirmary's desire to set up an NHS Department of Haematology but that he had received no reply. At this stage I have to explain that the "University" is basically a corporate body of Professors of whom each Professor has absolute power within his or her Department. Professors when in negotiation would say "I agree" when in fact they did, but if they did not agree then they would make some comment like "the University will not like this", or "I doubt whether the University will approve of this proposal". So when Arnold Tunstall wrote his letter to the University it would have been despatched straight to Lumsden who would have put it in the bin. As a member of staff of the University there was no one who could be approached above the level of Professor if there was a particular problem. I would then from time to time make an appointment and see Arnold Tunstall to see if there was any progress only to find out that further letters had been written and again there was no reply. This situation went on for well over two years.

In the meantime I thought that I should be prosecuting some research and improve my CV. It was obvious from the routine I have just described that it would be difficult to pursue a serious research project. At the time there was considerable interest in platelet function and Geoffrey had set up techniques for determining platelet aggregation and this was a subject I would have liked to have followed. My programme of work, however, forbade this.

It seemed therefore that I should look for a project involving morphology which could be activated at any time. I attended the Friday Clinic as often as I possibly could and had a major interest in polycythaemia vera and the myeloproliferative disorders. Colin Woods, the Bone Pathologist in the Department of Pathology had a laboratory in the Orthopaedic Department which was producing some interesting work. He had a series of point array graticules which were placed inside the microscope and produced a pattern of dots across the microscope image enabling one to pursue a counting technique to measure such things as the width of osteoid seams. I decided therefore to make a bone marrow aspirate and trephine examination of patients with polycythaemia vera and myelofibrosis to try and describe the evolution of myelofibrosis in the polycythaemia vera. Colin Woods also taught me to use polarised light in the examination of a trephine biopsy which was extremely helpful.

I decided to write an MD on this project but was very careful not to let Professor Lumsden know that I was undertaking this or he would have found means to block it. The Friday Clinic was run by

Dr David Miles, a Lecturer in the Department of Medicine, and Jim Fountain popped in from Keighley from time to time. I found myself giving the radio active phosphorous on a regular basis and this continued until the last Friday before I retired.

At this time I encountered a random collection of haematological cases which were often pancytopenic, and morphological abnormalities of the neutrophils and, on some occasions, evolved into acute leukaemia. I performed a review of the literature and found out there were such names as idiopathic pancytopenia, chronic refractory anaemia with sideroblastic bone marrow, and a pre leukaemic acute leukaemia. I tended to favour the term pre leukaemia but found difficulty in explaining this term to patients.

I went on to write up four cases with pre leukaemia, a very small number but there was very little in the literature at the time. This topic went on to be a major interest of mine for the rest of my career. A further problem with this diagnosis was the interest at the Hammersmith Hospital, in particular Professor Mollin, in sideroblastic anaemia. There is no doubt that idiopathic sideroblastic anaemia is a pre leukaemic disorder but it was regarded by Mollin and his colleagues at the Hammersmith as a nutritional problem and they set out to find the missing nutritional factor which was, they thought, an analogue of pyridoxine. This went on until Professor Mollin moved to St Bartholomew's Hospital where Dr Child, my future consultant colleague, obtained an MRC research post specifically to find this missing nutritional factor and, which they thought, was akin to the search for Vitamin B12 in pernicious anaemia. Of course they never found it.

Further reports appeared in the literature and I found that the term myelodysplastic disorder, which describes the predominant morphological feature, could be the best descriptive term for this disorder. However regrettably in later years the FAB (French, American, British) team made the decision to call the myelodysplastic anaemias, refractory anaemias. This used a term going back many years to patients with a raised MCV which did not respond to B12 or Folate. I found this term very retrogressive and in reporting cases I found it very difficult to use. Not all sideroblastic anaemias however were myelodysplastic. Monty Losowsky, subsequently Professor of Medicine at St James's, found a very large kindred of hereditary sideroblastic anaemia which he pursued by travelling all over Great Britain in his car. When I was at the Hammersmith I met him there discussing the case of a family with Professor Dacie.

Alas as time passed by there was still no reply from the University but however I was still very determined to see this through. New haematology posts however were being created through the country and it was particularly difficult not to apply for some of these. I had formed a Regional blood club which met from time to time on Saturday afternoons in the Infirmary. This was proceeded by an excellent buffet lunch with whole cold salmon and legs of boiled ham which were provided by the Infirmary Endowment Fund. At these meetings Lionel Sacker, who was a Consultant Haematologist at Hull, put me under considerable pressure and called me an idiot for hanging around.

In 1968 we appointed a new technician by the name of Philip Day. He had left school at the age of 17 and, according to his parents, was destined for a job in the office of Dewsbury City Council Cleansing Department. By appointing him we rescued him from this fate. He was to make many significant contributions to the development of the Department. Philip was a highly intelligent young man who should have gone to University and done an Honours Degree Course in Mechanical and Electrical Engineering, but he chose not to do this and probably received poor advice at school. He developed and took responsibility for developing and performing the tests connected with the haemoglobinopathies and red cell biochemistry problems. He also took an active interest in routine haematology and I was very pleased when under my instigation he took a serious look at the staining of blood films. My major skill was in morphology and it was absolutely essential that I had slides stained to perfection on every occasion. The first problem with making blood films is that if cheap glass is used it is very alkaline and alters the stain. We had first of all therefore to look for various suppliers until we found the ideal one, although it had not to be too expensive or Professor Lumsden would have blocked the order. It also determined that there was no alternative to buy Analar grade of alcohol to fix the films. If this was not used the red cells appeared to develop bubbles due to a content of water. Another problem was that the stain May Grunwald Geimsa could be variable and the same manufacturer would send variable batches of stain. Philip got to work on this and appreciated that Geimsa was a mixture of azure blue and eosin. We began to order these two reagents separately and mix them in the laboratory but this turned out to be prohibitively expensive. In the end we did find a manufacturer that produced reliable and consistent Geimsa stain. Philip tells me that on one occasion however there was a disaster. He and one of his colleagues, Raymond, went down the steps to the cellar where the reagents were stored and Raymond dropped the tin of Geimsa. The lid flew off and a fine purple powder emanated from the tin into the store room and Philip and Raymond and all their clothing developed a beautiful mottled purple stain.

In 1969 I went and met with the Priorities Committee once more and asked for a new automated cell counter that had just appeared on the market and which had revolutionised haematology. It was the Coulter Model S and cost £100,000. Much to my delight the Priorities Committee agreed to purchase the piece of equipment to be delivered the next year. I went to see Arnold Tunstall and he told me that he had written to the University but this time had said that if he received no reply to this letter then it would be assumed the University agreed and the Infirmary would go ahead.

In February 1970 the Coulter S was delivered and for Philip Day it was love at first sight [page 128]. It would perform 90 blood counts a minute and would accurately count the red cells and measure mean cell volume. These were dramatic advances., There is a condition known as thalassaemia minor in which there are hypochromic red cells and to compensate for this a high red cell count. I had stated that I had never seen a case of thalassaemia minor in Leeds at this stage but within the first morning of use we saw our first case because of the performance of the Cell Counter. What was very interesting was that Professor Lumsden performed one of his very rare visits to the laboratory on the morning that this was delivered. He of course knew absolutely nothing about this and made no further comment.

The interview for the new Consultant post in Haematology was organised for June 1970, the day before the General Election. The Chairman of the Board of Governors, Sir Donald Kaberry, agreed to chair the interview expecting the process to be a short one but it turned out not to be so. There was one other applicant besides myself. The interview itself was unremarkable. However the wait afterwards for the Committee's decision went on and on for at least two hours. I was given a blow by blow description by Colin Geary a good friend of mine whom I met when we took the MRC Path exam together in Glasgow and who was the representative of the Royal College of Pathologists. In fact, Arnold Tunstall had asked the Royal College of Physicians to provide a representative. When his error was pointed out to him at the interview he was however able to ring the Royal College of Pathologists and ask them to accept Dr Geary as he was a member of both colleges.

The main problem at the meeting was the fact that Professor Lumsden was doing all he could to wreck it and at the same time Sir Donald was getting more and more irate as he was missing out on last minute electioneering. Finally Professor Lumsden decided to play his ace of trumps. He stated that a Consultant Haematologist should have an allocation of beds and clinical facilities and there were none. However Professor Sir Ronald Tunbridge who was at the meeting answered this by saying that he would provide beds on the Professorial Medical Unit. This led to the conclusion of the meeting which resulted in my appointment as Consultant Haematologist and the fact that Professor Tunbridge had offered clinical facilities saved me the bother of a lot of negotiations once appointed.

In fact, Professor Lumsden had done me a great favour in the end. Thus I was free from the clutches of Professor Lumsden for the first time for many years and I have long since wondered why he appeared so afraid to relinquish control of Haematology. On weighing up the evidence over the years, including the virtual dismissal of John MacIver, the fact that I was told on one occasion that there were two haematology technicians in his research unit, which I did not believe at the time, and the fact that I met a Consultant Haematologist who had worked in Lumsden's research unit when on an inspection of a hospital with the Joint Committee of Higher Medical Training, and he told me that all members of Professor Lumsden's research team in Clarendon Road had been told most emphatically that they must never ever speak to Roberts. All this meant to me that Professor Lumsden was partly funding his research unit out of monies received from the NHS to fund Haematology. This of course was completely against any agreement between the University and the NHS over funding but there was no control whatsoever of professorial spending in those days. So I was now free from Professor Lumsden. I never visited the post mortem room again and looked forward to my future career with enthusiasm, but if I thought I was rid of Scottish professors, then I had another think coming.

CHAPTER 7

THE PROMISED LAND

The day after I was appointed Consultant there was a general election. The time preceding the election the Chancellor of the Exchequer, Roy Jenkins, had been applying cuts within the economy which proved unpopular and just before the election there were unfavourable economic figures. This lead to the election of a Conservative government under Edward Heath. Edward Heath and his Chancellor of the Exchequer, Tony Barber, decided to pursue a different course in the economy altogether. They went for growth and the result was called "Barber's boom". This was a failure and at the next election Labour were returned. There then followed a very significant and radical reorganisation of the National Health Service lead by the Secretary of State, Barbara Castle, and the Minister of Health, David Owen. This would have a very significant effect upon the General Infirmary at Leeds. There then followed very significant Union activity which was leading the country into ruin. There were many industrial strikes and Barbara Castle, in a new ministerial post, drew up a document called "In Place of Strife". This, again, was ineffective against the ever increasing power of the Unions. There then followed a very turbulent period in this country.

Heath was re-elected but nevertheless there followed a three-day week, a miner's strike, and subsequently on re-election of Labour, under the premiership of Jim Callaghan, there then followed a winter of discontent when bodies lay unburied and shop stewards drew up operating lists in the National Health Service according to their priorities. This lead to a further general election in 1979 when the Unions were finally confronted by the "Iron Lady", Margaret Thatcher. This did not mean peace but further turbulence and an all out miner's strike.

It was against this background that the Infirmary hoped to acquire many millions of pounds to build a new hospital. Planning for this new hospital began after the war and it was proposed that the entire Infirmary, Medical School, and Hospital for Women complex be raised to ground and a brand new hospital built. This was further modified on appreciation of the fact that the Medical School and Infirmary were listed buildings, so that the Old Medical School was preserved and the front façade of the Infirmary retained.

When I was appointed the planning process was in full operation with an anticipated opening of the new hospital in 1984. There were many Orwellian jokes and comments about this particular date.

I was soon involved in this planning process and seemed to spend many hours drawing up plans for a brand new Department of Haematology. My view was that the hospital would never be built and that this planning process was but a waste of time but I co-operated with a planning team which comprised some thirty people. I had planned a superb new department approaching the standard now obtained in the Bexley Wing and asked for a further eight consultant posts. These requests were not challenged and I felt that it was just a matter of time before the whole project was closed down.

The last seven years in the University Department of Pathology were very difficult and building a Haematology Department a struggle. However by the time I was appointed a Consultant I had an excellent team of Medical Laboratory Scientific Officers (MLSO's) lead by the Senior Chief, Charles Buchan. Philip Day, who had developed a complete mastery of the Coulter Model S, had a great interest in red cell problems and, together with Colin Toothill from the University Department of Chemical Pathology, he developed a wide range of red cell enzyme assays and an excellent diagnostic service for the haemoglobinopathies. Geoffrey Tate enthusiastically developed the diagnostic service in coagulation and took a particular interest in the lupus anticoagulant. If there was a coagulation problem in the Laboratory he would proceed to the ward, take further samples, and would have usually worked out the diagnosis by the time a member of the medical staff came to see the patient. Interest in the lupus anticoagulant resulted in him working in close collaboration with John Turney, the Renal Physician and the Professor of Obstetrics, James Scott. Kathy undertook the microbiological assays for folate and went on to develop the radio-immune assays for serum B12 and folate. The rest of the Department were highly motivated and skilled and remained as good looking as ever.

The major problem in the development of the Laboratory was that of out-of-hours duties. This was voluntary but very remunerative. It was however also very demanding in that the technician had to work one whole day, then do on-call overnight, and work the following day. As the work got more and more busy and more demanding then the technicians got more and more tired. Moreover many placed their on-call remuneration in their mortgage applications and were therefore trapped to the on-call system.

The work in general continued throughout the years as numbers increased. I put the annual statistics on semi log paper and showed that the rise in blood counts was exponential and this rise continued until I retired. There were problems in dealing with this progressive increase and at one stage Professor Lathe in Chemical Pathology introduced rationing tickets for individual firms. This system however soon failed and it appeared there was no solution to this problem apart from ever increasing rationalisation and automation.

It was finally decided to progress with Phase 1 of the new hospital. This entailed knocking down the Hospital for Women which was immediately adjacent to the Algernon Firth Institute of Pathology and transfer the work to Roundhay Hall which was formerly used for nurse training. There was a major problem however and that was that it would not be possible to move the Blood Bank to Roundhay. It had to remain on the Infirmary site and the planners had no idea what to do about this problem. As a new Consultant they approached me and I suggested that it would be possible to modify the present Haematology Laboratory to accommodate Blood Transfusion. When I was a student there used to be a post mortem demonstration to the entire medical year of some sixty students and the PM room then was, in part, a tiered amphitheatre with a Pathologist in the centre dissecting the body. The scene resembled one of the Renaissance paintings of the early anatomists. By the time I was qualified this had been removed and the PM room reorganised. However the steps to this amphitheatre were still present as was the ante-chamber which was not used. I proposed that the remaining flight of steps would be an ideal way to provide an upstairs floor in the Department of Haematology. The planners took up this suggestion and work began.

The ground floor was upgraded to accommodate a Transfusion Laboratory. Upstairs there would be a bedroom, particularly for the Blood Transfusion technician to sleep in, a Registrar's room, a room for a new Consultant, toilets, and a room for me which was large enough to hold several people for teaching purposes. I had installed a microscope which projected with light from a carbon arc onto a large metallic conclave screen, and also blackout curtains. Thus we were able to have morphology meetings. The secretary's room adjacent to mine was small and would accommodate one secretary. However, we soon acquired after this a filing clerk to deal with all the new stationery needed in the treatment of leukaemia and for the two ladies appointed the office became a battle ground. One could not abide heat and the other cold, so inevitably when I left my office I would often walk over a prostrate secretary either switching the central heating on or off.

The Infirmary was directly managed by the Department of Health and a considerable amount of planning and spending freedom given. Thus I was able to get a whole range of patterns of carpets to look at and by the time the new upstairs had been constructed and carpeted it looked very attractive and luxurious. It was subject to many sarcastic and barbed comments over the degree of luxury that we lived in, this however would never be repeated again.

The Department of Blood Transfusion therefore moved over onto the Infirmary site with dramatic effect. The Senior Chief Technician in charge was Mr Ken Major who was a stickler for high standards and something of a martinet. He totally re-organised the use of blood within the Infirmary. He arranged for a new Blood Transfusion fridge in Theatre and others in strategic points in the Infirmary all of which had an alarm which was connected to the telephone exchange. The storage of Blood Transfusion bags in the ward fridges was forbidden. I had no worries whatsoever about the Blood Transfusion Department as it was run to a very high standard.

Ken Major however did present me with problems from time to time in that he set specific rules in Blood Transfusion which had to be followed and he expected all the clinical staff on the wards to similarly obey rules that he had laid down but this did not work. There was a continuous problem of specimens of a routine nature arriving late in the afternoon for blood to be cross-matched for use in Theatre the following morning. The specimen would arrive too late for the day staff and too many to be coped with by the on-call staff. I was repeatedly requested by Ken Major therefore to go and give a good dressing down to the Professor of Obstetrics, Professor of Surgery, and Consultants performing open heart surgery. This was a problem however that never ever seemed to go away. The Consultant in charge of Blood Transfusion however was Dr Alan Ambery-Smith, a Gynaecological Pathologist. Ambery-Smith was formerly a Lecturer in the University Department of Obstetrics and a very good teacher and obstetrician. He failed however with the lack of expansion in the National Health Service to get a Consultant post and was placed in the Department of Pathology in the Hospital for Women to learn gynaecological pathology from the existing pathologist.

My next immediate job was to organise junior medical staff. I knew one of the registrar posts belonging to the University Department of Pathology was in fact, right from the beginning, a haematology post funded by the Infirmary and filled first of all by Arthur Bloom as a Registrar in Haematology and then by me. This post I immediately claimed. There was also very often a rotational Registrar from the Pathology Training Scheme of pathologists from various disciplines, as was required by the regulations of the Royal College of Pathologists. Shortly after appointment I received a telephone call from Arnold Tunstall, the Hospital Secretary, asking if I would see him. Arnold told me that he had received a letter from the Department of Health advising him that we had been awarded two Senior Registrars in Haematology. I was highly delighted by this news and went to see James Lynch, the Post Graduate Dean. He told me however that he had had a copy of this letter and had written to the Department of Health turning both posts down. James Lynch explained that there were nine posts in Pathology, emphasising the word Pathology, and that only three of these were filled. He said therefore that one could hardly expect to take on two further vacancies. I asked him then to advertise two of the existing vacancies as Senior Registrars in Haematology, but he then explained to me that this would not be possible. Senior Registrar posts in Pathology were advertised as Senior Registrars in Pathology and then the successful candidate at interview would declare his major interest. The whole of West Yorkshire therefore still embraced Clinical Pathology and there appeared to be no mechanism for advertising for posts in specific disciplines.

However, James relented and I was soon able to appoint a Senior Registrar in Haematology but the position of Haematology as a speciality was soon helped when the College allowed the examination of Membership of the Royal College of Physicians to exempt the trainee from the primary examination of the Royal College of Pathologists. This step transformed the picture in Haematology and we were soon appointing Registrars with no previous training in Pathology who had been Registrars in General Medicine and received a thorough clinical training.

One of these early posts was filled by Dr Ian Burton who had been formerly Registrar to Professor Losowsky at St James's. The post that Ian Burton filled was a regional one and he had therefore to undergo rotational training. He therefore rotated to Bradford Royal Infirmary to receive further training in Haematology.

On his arrival the first morning he was approached by Dr Bert Kellett, the Senior Pathologist, who said to him "Burton, I have two post mortems for you this morning." Burton replied that he had never performed post mortems and was very unlikely to do so. Bert Kellett was aghast. The scene at Bradford was reminiscent of an H

M Bateman cartoon. Bert Kellett retired to his room and the telephone line to the Regional Health Authority was soon incandescent, but no matter how much he expressed his rage it finally dawned on the Pathologists of Yorkshire and the Regional Health Authority that Haematology as a single specialty had arrived.

The discipline of Haematology was now about to undertake a major revolution. New agents for the treatment of leukaemia, such as, vincristine, daunorubicin, cytosine arabinoside, 6-thioguanine, had recently been introduced and were being used with the established drugs, such as, 6-mercatopurine, methotrexate and steroids.

It was however the pioneering work of Skipper in the United States that set the scene for the modern treatment of haematological malignant disorders. About this time Skipper produced some original work on leukaemia in mice. He inoculated susceptible mice with a leukaemia cell and estimated that it grew exponentially until it killed the mouse. He proposed that growth of leukaemia cells was exponential and killed the mouse when there were 10^{13} cells in the body. This is not strictly true because, of course, there is inevitable cell death in any system of cell proliferation. However the concept was useful and Skipper produced charts to show that when a patient relapsed there were 10^{12} logs of cells and when the patient died there were 10^{13} logs of cells. A very interesting concept was the fact that when a patient went into clinical remission there were still 10^{10} cells or 10 logs of cells, that meant that the treatment of leukaemia had to continue long after there was apparent remission in the patient.. Skipper went on to demonstrate that chemotherapy needs to be given at intervals. This is based on the fact that on administration of therapy there is cell death of leukaemia cells and normal marrow cells, but that normal marrow cells regenerate more quickly. After every course of therapy therefore there is progressive reduction in leukaemia cells and a progressive increase in normal marrow cells. It was also established at the time that chemotherapy drugs are best given in combination. Thus the theoretical pattern of the treatment of acute leukaemia with drugs used in combination at intervals was established.

To inform the Infirmary medical staff about what we were likely to be getting up to I invited Professor Gordon Hamilton-Fairley, from St Bartholomew's Hospital in London, to speak to us and tell us of his successes in treating leukaemia and lymphoma at Bart's. I found the lecture very interesting but was disappointed to hear comments when walking on my way out of the lecture, such as, "what a load of rubbish, and I don't believe a word of it". But they were to be proved dramatically wrong.

Changes were taking place with the staffing of the Friday Clinic. For two or three years David Miles, a Lecturer in the University Department of Medicine, had been running the Clinic with help from me, and from time to time Jim Fountain would pop in. David Miles obtained a Consultant post at the Airedale District General Hospital, and I received a letter from Jim Fountain telling me that in view of all the changes that had taken place he was now resigning. Therefore I now had the Clinic to myself
.At this time patients with malignant disease were not told their diagnosis.Relatives were fully informed but patients kept in blissful ignorance.

I inherited, from Jim Fountain, a patient with chronic myeloid leukaemia. On his notes were express instructions,from his parents,that under no circumstances must he be told the diagnosis.The patient presented on honeymoon with a haemorragic syndrome due to the fact that he was hypersensitive to busulphan, the standard treatment for chronic myeloid leukaemia.Patients who react in this way often die of marrow failure but some slowly regenerate and may live for 20 years or more.

Unfortunely the patient's marriage failed and he was divorced.There then followed a custody battle for their son.The patient then asked me if there was any thing he should know before the court hearing.I had no option but to tell him his diagnosis.The patient was distressed and his parents furious.

Shortly after this a patient in Glasgow was treated for chronic myeloid leukaemia with busulphan and reacted conventially but not told the diagnosis.He felt well,went to see his bank manager,borrowed a large sum of money and started a new business. He then, accidentally,found out his diagnosis,went beserk,and smashed up his home and had to be restrained by the police.

This story appeared in all the national newspapers and had a profound influence on attitudes of the medical profession.It became clear that the doctor had a responsibility to tell the patient the absolute truth about the diagnosis and outlook

As I have already mentioned, as a result of the disastrous MRC Trial in the early 60's, Sister Freda Ellis, and Sister Rita Cox, were completely opposed to further MRC Trials in leukaemia. Patients however did turn up on the ward and were referred to us with acute leukaemia so they had to be treated. Both Freda and Rita exhibited the opiates to help the patients to a better world but then one day a young man, a student at the Metropolitan University, was admitted with acute leukaemia. He was treated with a regimen based on the work coming out of Bart's and the Royal Marsden Hospital. The patient was rendered completely pancytopenic but then on one day I noticed a little flicker upwards in the platelet count. This continued and the patient went into complete remission. The student was therefore the first patient with acute myeloid leukaemia in Leeds to achieve complete remission and he literally took up his bed and walked. He remained in remission for several months until inevitably he relapsed.

I rejoiced at this and Freda and Rita displayed their amazement because they never believed this could possibly happen. Suitably impressed therefore we continued to treat more acute leukaemias as they came in. Management was not easy however and remission rate low because there were no intravenous antibiotics and no platelet support. In fact, the main advances in the treatment of acute leukaemia in the years to follow were in the supportive therapy, that is keeping the patient alive whilst the chemotherapy worked.

We began to see therefore patients in remission with acute leukaemia for maintenance therapy, as well as lymphoma patients. It was clear to me that management in Out Patients in a conventional sense, that is patients attending the Clinic declaring how they were and then looking at the blood count and writing prescriptions, would not work because of the time involved. I therefore devised a system for Out Patients with the help of Mike Galvin who had joined us as a Registrar in Haematology from a similar post with the Professorial Medical Unit. The Professorial Medical Unit was particularly interested in what was known as problem orientated medicine and Mike Galvin brought ideas from this system of note recording. I had already organised the blood counts to be recorded on the graphs that David Galton had introduced at the Hammersmith which I found particularly helpful. What I did first of all was to organise a pre-clinic meeting.

Present were myself, a member of the junior medical staff to write prescriptions, often a member of the Pharmacy, a member of the Blood Transfusion Unit, and a secretary. Each patient was then reviewed, their place in the treatment schedule noted, after that the prescription for the forthcoming visit to clinic could be written. An estimate was made of the possible need of blood transfusion so the Blood Transfusion Department would know which samples were likely to be coming along and also determine whether any special blood was required if, for example, the patient had developed antibodies. Then request forms were filled in for other investigations, such as, chemical pathology and immunology. Thus when the patient came to clinic the next day all samples would have been taken, there would be no need to return to the phlebotomy suite for further samples, and a quick telephone call to Pharmacy and Blood Transfusion informed them of the need to send the appropriate therapy to Out Patients for treatment, or whether blood transfusion would be needed. In terms of the problem of problem orientated medical recording we arranged that there would be a common pattern of dictation which included, in order, the following:
S - subject - how the patient feels, and symptoms.
O - objective - the blood count and any other investigations that had been performed, or physical signs.
A - assessment - particularly of where they are in the chemotherapy schedule, and
P - what is planned to do.

This would very often say to give the therapy and what it was today and to return in two weeks time and then the drugs and dosage that would be given.

Thus the clinics began to run quite easily and the patients did not have to hang around very long. One major problem was the lack of a One Day Unit because more and more patients, particularly those with myelodysplastic syndromes, required transfusion. They had to be got onto the ward and sit anywhere, often on a bed vacated by a patient, to have a transfusion. Over the years I continuously requested that there should be a One Day Unit but this was vehemently opposed by the nursing staff on the grounds that the patients would inevitably stay in after 5.00 pm and therefore have to be admitted. My point was that the patients should only be transfused that amount of blood which would guarantee discharge before 5.00 pm. I had previously had words with Martin Israels in Manchester who had explained to me why there was never any reason for a patient in a One Day Unit to stay longer than 5.00 pm providing the system worked well. The logic behind this is that if a patient has an aregenerative marrow they make no red cells. Red cell life is approximately 120 days therefore approximately 1% of red cells are destroyed each day and therefore have to be replaced. This means that if the red cells are not replaced the haemoglobin will fall by 1 gm each week. Every patient has one unit of blood then the haemoglobin usually rises by 1 gm. I thought that if the patient is on regular blood transfusion treatment for an aregenerative bone marrow they need one unit a week so they should return every two weeks for a two unit transfusion which can be given quickly. If however the patient returns,for example, after five weeks and is in anaemic heart failure then inevitably they will have to stay in while the transfusion is given very slowly.

The chemotherapy drugs were sent down to Out Patients where I had to roll up my sleeves and make up the treatment by adding diluents to various bottles and giving them either by injection or into a free flowing drip. At the time the Out Patient Sister was Sister Carol Bilborough who took an interest in what I was doing and before long got herself a rubber pinafore and rubber gloves and took this task away from me. Little did she know when she took this over that she had committed the rest of her working life to the Department of Haematology and in fact is still continuing to work as the Secretary of the Friends of the Leukaemia & Lymphoma Unit.

For the first six months that the Haematology Clinic was working under the Department of Haematology we also had to look after paediatric leukaemia because there was no one in Leeds with any experience in this field whatsoever. We saw several patients with acute leukaemia, most of them went into remission on an Out Patient basis but subsequently relapsed. Some of them had a central nervous system relapse which very often manifested itself as the so-call hypothalamic syndrome in which the children become very hungry and eat enormous quantities of food and rapidly become obese. For these children it was I am afraid a death warrant. However in the course of this six months we did treat a little girl with acute lymphoblastic leukaemia who became a long term survivor. The treatment of children with acute leukaemia was finally taken over by Dr Katherine Howarth, Paediatrician, who set up an appropriate clinic and was allocated some beds. I remember meeting her in despair one day as she had been away for a week's holiday and came back to find many of her patients on the wards septicaemic, but untreated, because the Paediatricians refused to give antibiotics until the bacteriology reports came back from the laboratory. It was standard practice in acute leukaemia that when patients become pyrexic they are treated immediately with antibiotics otherwise they may rapidly die. Eventually Katherine, or Katy as she was known, took a sabbatical to go work with Pinkel in the United States who had introduced central nervous system prophylaxis in childhood leukaemia by irradiating the skull and spinal column and giving intrathecal methotrexate. Katy did not however return. It is rumoured that she fell in love with Pinkel and decided that her future was with him.

Following her resignation Dr Clifford Bailey was appointed as a Paediatric Oncologist to look after the Paediatric Oncology Unit at

Seacroft Hospital. Under the management of Cliff Bailey the Unit developed as a Paediatric Oncology Unit of high repute.

As the Friday Clinic became busier it was obvious that we needed a specialised Lymphoma Clinic. I approached Dr Geoffrey Stone, Consultant Radiotherapist, to get his collaboration in the setting up of such a Clinic. A room and nursing staff were allocated to us for a Lymphoma Clinic to be held on a Tuesday morning. However there were several Radiotherapists at Cookridge Hospital, and others, who were interested in such a Clinic and so a joint meeting was held in the Lymphoma Clinic to discuss its future organisation. At the first meeting there were Professor Kunkel, Head of the Radiotherapy Unit at Cookridge and Dr Stanley Worthy who was a Consultant Radiotherapist at Bradford who had a major interest in lymphoma. He was very keen for us to join the National Lymphoma Investigation (NLI) which was a therapeutic trial developed by and run by the Royal Marsden Hospital. Also from Cookridge was Dr Lorne Campbell-Robson who, I think, attended because of general interest and curiosity. Professor Teddy Cooper, Professor of Cancer Research, also came and wished to participate. There then followed endless dispute between Stan Worthy and Geoffrey Stone. Geoffrey would not accept the protocol for the NLI because he said that there was a risk of over lapping fields in the scheme for radiotherapy treatment. On this point he refused to relent. The atmosphere was very argumentative and Peter Kunkel, who was an extremely pleasant, calm, and rational man, managed to keep things under control.

I remember on one particular Tuesday that Dr Rex Tattersall, Consultant Physician, had the adjoining clinic suite and walked in with a patient with non-Hodgkin's lymphoma. His face was a picture as he presented one patient to six consultants. A history of the patient was obtained and each took it in turn to examine the patient, so at least the Clinic had started.

Peter Kunkel however, a great diplomatist, shocked us all by announcing that he was giving up medicine and was going to run his father-in-law's farm in Devon. From this time onwards relations began to deteriorate and it was not very long before Geoffrey Stone and I were left to run the Clinic.

We began to receive referrals and obtain remissions and the Clinic became a great success. Geoffrey Stone however was not an easy person to deal with. He was a complete obsessional and looked at every patient's notes with incredible detail less he should miss a single point or blood count. Notwithstanding this the treatment the patients obtained was meticulous and I would have accepted him personally as a Consultant any time. He would, however, turn up at National meetings and in the middle of the meeting stand up and start to make several points to considerable groans from around the audience. The point was however that Geoffrey was inevitably correct in what he said and had picked up mistakes in the protocol which everyone else had missed. On the personal side Geoffrey was an accomplished violinist and would amuse himself in the evenings by playing his violin with a recorded accompaniment. I remember on one occasion he had attended a concert at which Yehudi Menuhin played and triumphantly told us that he had personally spotted several mistakes .

When Geoffrey retired I am afraid he did not live very long and had several strokes. At his funeral however I was amazed to hear that when Geoffrey was at school he was a very successful wing three-quarter for the rugby union team and represented the school in national athletic competitions in the 440 yards event in his native South Africa.

We had just established ourselves as a unit with Junior Staff, beds and increasingly experienced nursing staff when the Medical Research Council (MRC) announced that they had reinstated the Working Party for Adult Leukaemia. The Chairman was Professor Sir John Dacie and the Secretary was Professor David Galton. The country was divided into regions and I was invited to represent the Yorkshire Region. It was not only my job to enter patients into trials but to promote them throughout the Region. Soon I was awarded a part time MRC clerk who attended all pre clinic meetings and kept all the forms and charts up to date.

This would soon be facilitated by the creation of a Regional organisation the Yorkshire Regional Cancer Organisation. The two main movers in this organisation were Professor Charles Joslin, Professor of Radiotherapy, and Professor Teddy Cooper, Professor of Cancer Research. The organisation was fully backed by the Regional Health Authority and was housed in temporary buildings at the Cookridge Radiotherapy Centre. There was also a full complement of administrative and secretarial staff.

The task of the organisation was to promote the effective diagnosis and treatment of cancer in the Yorkshire Region and to promote therapeutic trials. The MRC trials fitted in with this perfectly. One of my jobs as MRC representative was to hold meetings of Regional haematologists and promote the MRC trials. The YRCO helped in the organisation of this and provided a venue for the meeting. There was also a Blood Club to which Regional Consultants went and this met in various venues throughout the Region.

I also found myself in demand to give Post Graduate lectures on the treatment of leukaemia to doctors throughout the Region.

The meetings of the MRC Adult Working Party were held several times a year in the MRC Headquarters in Park Crescent, London, and were in themselves interesting and educational. The main trials to begin with were in Acute Myeloid Leukaemia led by Professor Frank Hayhoe and his team from Cambridge, and Myeloma led by Professor Jack Hobbs from the Westminster Hospital Trials - Chronic Lymphocytic Leukaemia and Polycythaemia Vera would follow.

Thus began a tradition for therapeutic trials in Leeds which has burgeoned with ever increasing numbers to this day.

At this stage the Laboratory and the clinical services were progressing satisfactory and I had ploughed a lonely furrow. This was however to change dramatically within the next year or two. I would be joined by Dr Tony Child the second Consultant

Haematologist, Professor George McNicol the Professor of Medicine who had a specialist interest in haemostasis and thrombosis, Professor Teddy Cooper who was a Professor of Cancer Research, Professor Colin Bird who had a major interest in the lymphomas, and finally Professor Ray Cartwright who took the post of the Leukaemia Research Fund Professor in Epidemiology.

The Dare haemoglobinometer. A film of blood between glass plates was compared with an inbuilt standard. The candle shown was for use in dim candle-lit or gas-lit rooms. The machine is being operated by Philip Day former Chief MLSO.

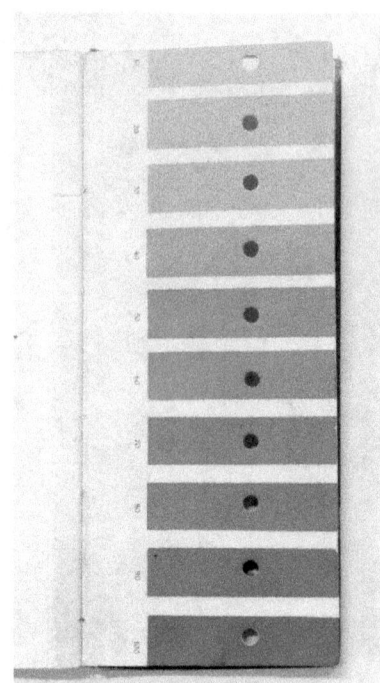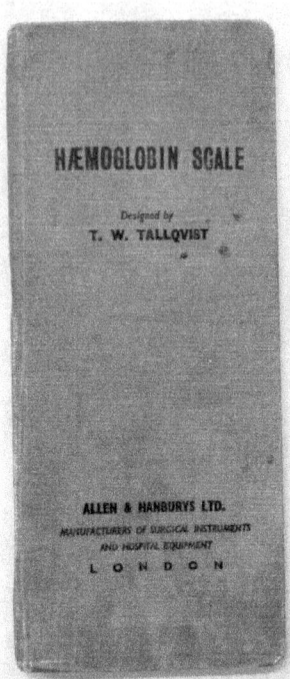

The Talquist card. (a) A drop of blood was placed in a piece of absorbent paper and passed behind the standard sheet which has holes in it to see the paper underneath. (b) These were produced in a book.

Sahli and Haldane equipment. With the Sahli system a sample of blood is mixed in a graduated tube with a fixed volume of hydrochloric acid to form acid haematin. Dilute hydrochloric acid is then added in drops until it matches the standard. The haemoglobin is then read off. With the Haldane system coal gas was bubbled through the diluted blood to form carbonmonoxyhaemoglobin. The acid haematin technique was widely used as a side-room technique.

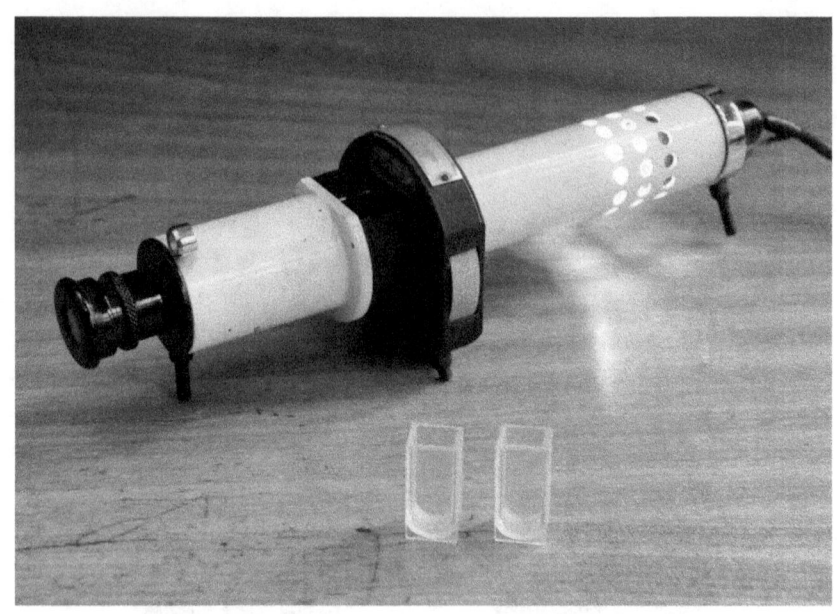

The Grey Wedge machine. Again a sample of diluted blood was placed in a cuvette and compared with a standard in the machine. The author used this machine on-call at the Hammersmith Hospital in 1963.

The EEL colorimeter. A sample of blood was diluted and converted into a stable pigment. The pigment cyanmethaemoglobin soon became the accepted standard. This was then measured in a colorimeter with a fixed wave length.

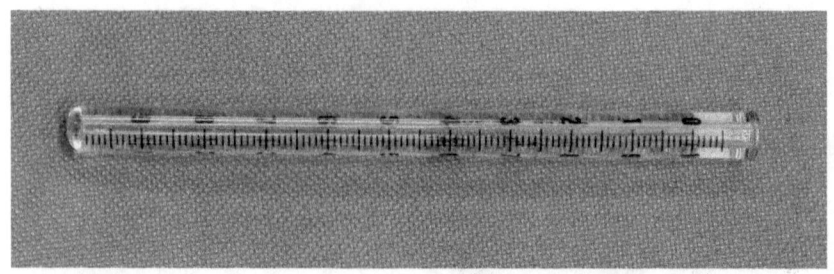

The Wintrobe haematocrit tube. Anticoagulated blood was placed in this graduated tube with a Pasteur pipette and centrifuged in a conventional bench top centrifuge to read off the packed cell volume (PCV).

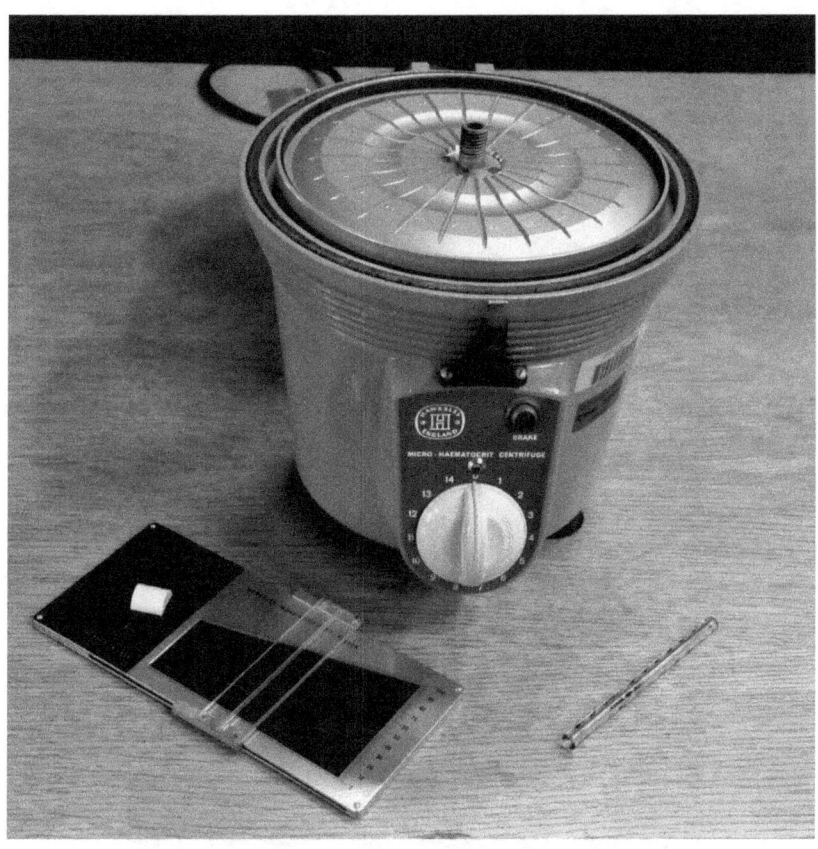

This was replaced by the micro PCV. A capillary tube was filled with sequestrine blood and subjected to high speed centrifugation in the machine shown. The PCV was then read in the specialised reader shown.

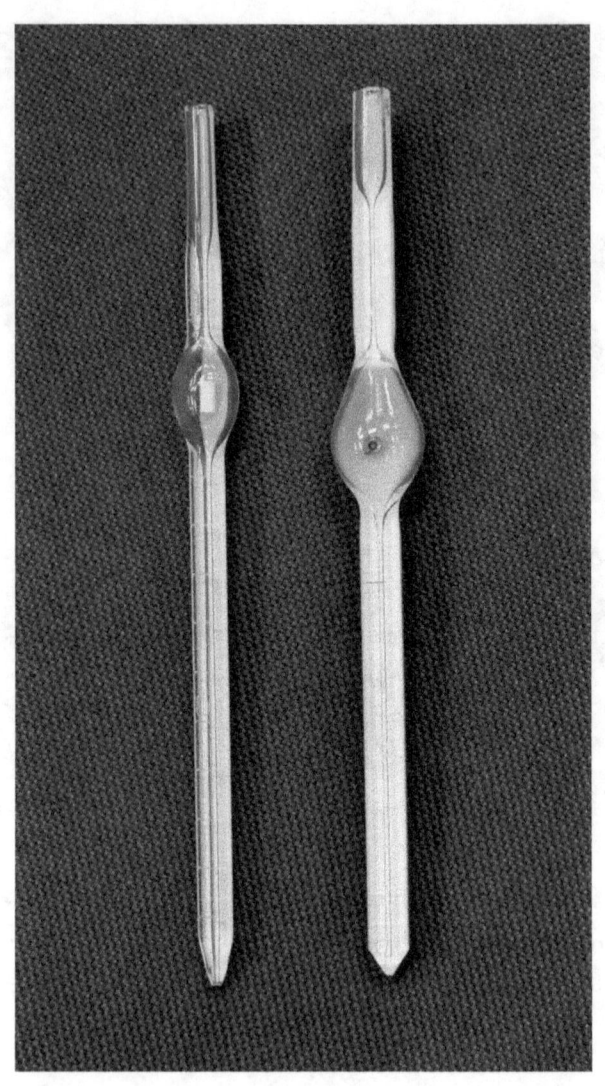

Two white cell counting pipettes are shown. The patient's finger would be pricked and blood sucked up by oral pipetting to a mark in the stem. Diluent would be sucked up past the bulb to another mark. The sample would be mixed in the bulb by rotating the bead in the bulb. The mixture would be read in a counting chamber.

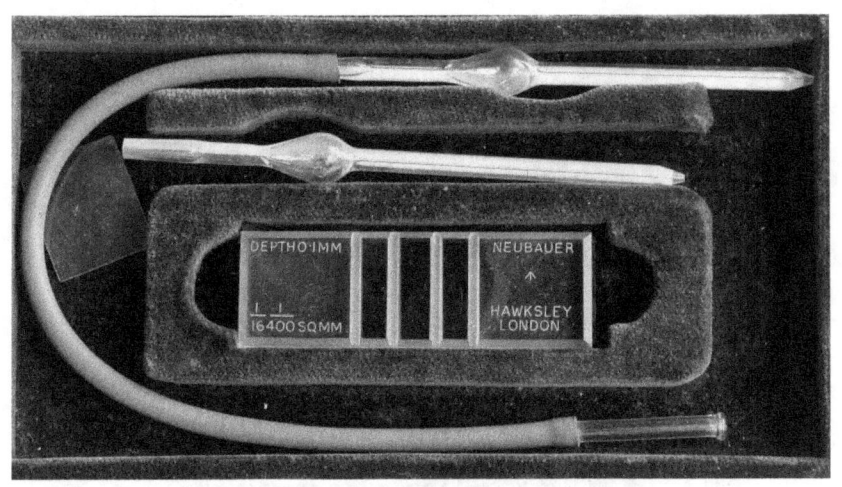

Pipettes and a Neubauer counting chamber.

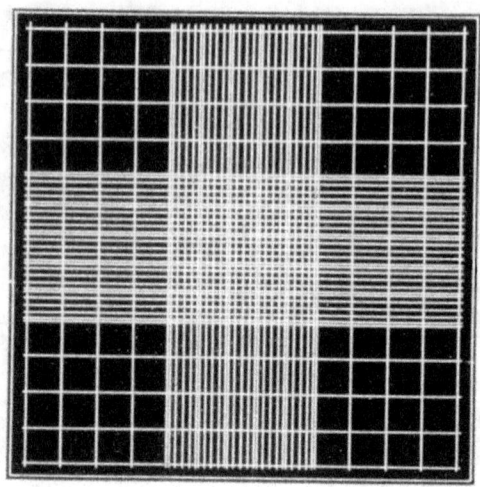

This is a view of a counting chamber seen through the microscope. The chamber had a cover slip and was constructed to contain a fixed volume of diluted cells from which a count could be made.

This is a mechanical counter used for differential white cell counts from a blood film. The clatter of this machine could be heard most of the time in the laboratory for many, many years.

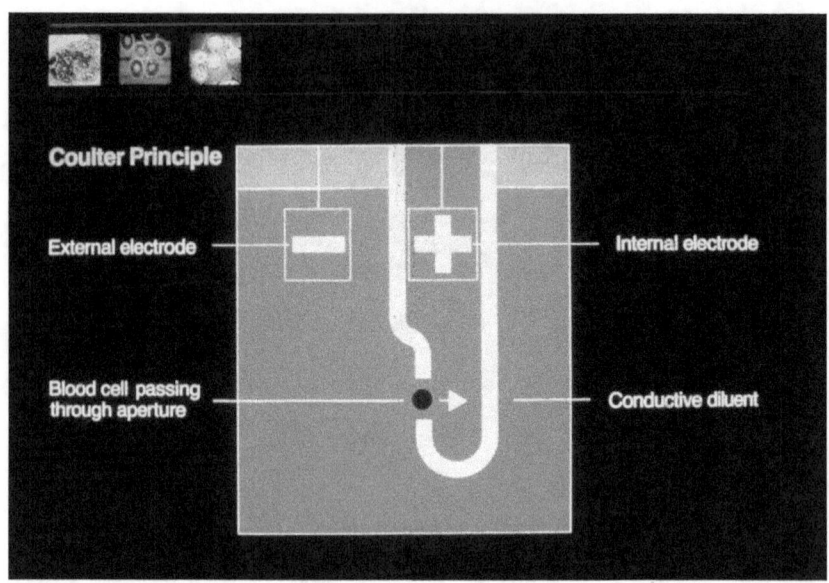

This shows the principal of the particle counter. Particles in a diluent are sucked through the orifice where an electrical current is interrupted every time a particle passes through. These disturbances are counted electronically.

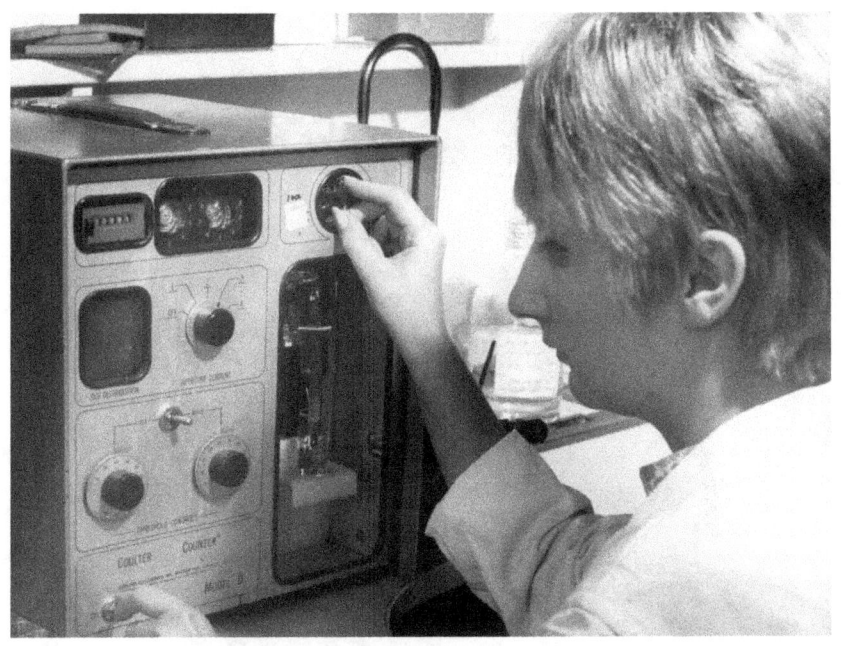

Jennifer using the Coulter Mode D which was used for white cell counts.

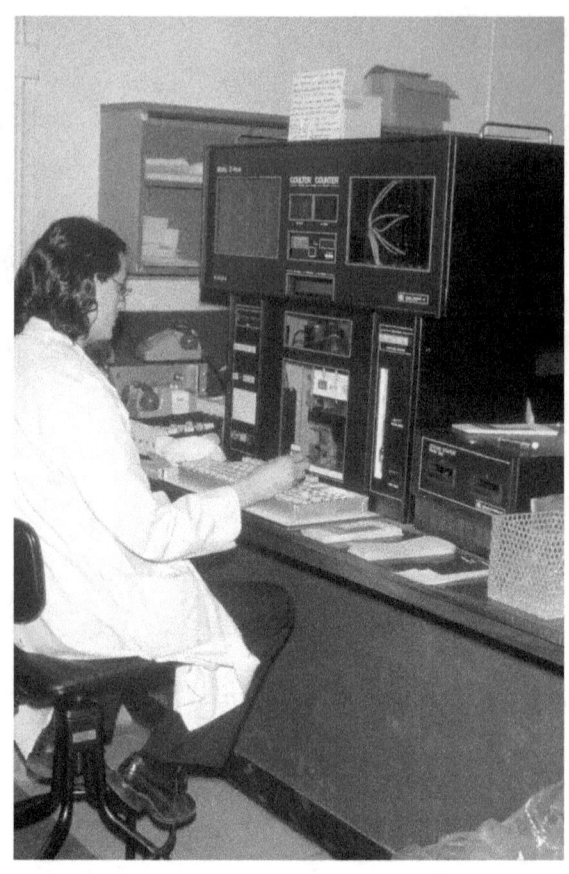

Peter Masters AKA "The Shad" using the Coulter Model S. This counter initially counted white cells and measured haemoglobin. It made direct measurements on red cell volume which was of great value. It printed the results on a report form and could produce 90 counts per hour.

CHAPTER 8

A POMPOSITY OF PROFESSORS

The Professor of Medicine

Clinical activity in the Professorial Medical Unit continued to progress with an increased number of referrals, some from regional hospitals. Most patients were entered into Medical Research Council trials. The Professorial Medical Unit junior medical staff were excellent and the nursing standards with Sisters Freda Ellis and Rita Cox were superlatively high. However, there was a cloud on the horizon. Professor Sir Ronald Tunbridge announced his retirement and the future of the haematology beds in the Professorial Medical Unit now seemed uncertain. The appointment of a new Professor was nothing to do with me and all I could do was await developments. I did observe one day George McNicol being entertained for lunch in the consultants' dining room, so obviously he was an applicant for the post. I had met George on one occasion for a chat at a railway station in a European country, following a scientific meeting and I knew his main interest was in haemostasis and thrombosis.

Some weeks later in 1971, the appointment of George P McNicol as Professor of Medicine was announced. When I met him he told me his major speciality would be haematology, particularly in the fields of haemostasis and thrombosis, both from the clinical and research point of view.

So it appeared that our interests would be complimentary and there would be no obvious clash. The same, however, could not be said for Dr Layinka Swinburne, who was Director of the Haemophilia Unit at St James's. George was, before his move to Leeds, Director of the Haemophilia Unit in Glasgow, one of the largest in the United Kingdom. George fully supported the presence of haematology beds in the Professorial Medical Unit and was keen to promote the future co-operation of the Department of Medicine and the Department of Haematology. I was overjoyed one year previously by the fact that I had got rid of one Scottish professor, but just over one year later I was associated with another, but this time within the NHS we were both of equal status.

George was appointed at a time of turbulence in the United Kingdom under the premiership of Edward Heath. In 1971 Heath was proposing to take the United Kingdom into the European Community. He was undertaking negotiations with President Pompidou and eventually reached agreement with him but there was enormous controversy in the United Kingdom and this finally led to a national referendum. In keeping with the fact that we were going to be fully fledged Europeans, Heath introduced decimalisation. Again this was hugely unpopular and through this introduction we were losing a currency which had been used since the era of the Anglo Saxons. This did have implications for medicine, we had to introduce SI units (Systeme International). This meant a tremendous amount of work at both national and local levels to convert the present units into the new ones and educate clinical staff. It is interesting to note that the Americans never have changed their units.

At this time the country was a very troubled one. There were uprisings of students who had revolution in mind and to this effect let off bombs here and there and this was followed by the resurgence of the IRA which began with Bloody Sunday in Belfast and was to go on and last years.

Heath also had to deal with a series of industrial strikes particularly with the miners and then in 1973 there followed the Yom Kippur war between Egypt and Israel which was a humiliation for the Arab world. The Arabs retaliated through OPEC the organisation of oil producers. They cut supplies and increased price. Then the miners decided to strike again. This ultimately led to the three-day week and petrol rationing was planned with the printing of petrol coupons and their distribution. Speed was cut by 20 mph to 50 mph on motorways. Heath finally decided that the country was ungovernable and went to see the Queen and asked for the dissolution of Parliament.

Against this background it was impossible to contemplate building a new Infirmary but a large area behind the Infirmary up to the University had been cleared, including the Hospital for Women, which had moved to Roundhay Hall. The Dental Hospital would move into the Worsley Building when it opened in 1979.

Amidst this almost inconceivable national turmoil Heath managed to radically reorganise Local Government and the National Health Service. Local Government reorganisation saw the disappearance of 800 councils and large new Local Authorities were created. The same principal was applied to the National Health Service with the creation of new area Health Authorities and Regional Health Authorities. This was an unmitigated disaster for the Infirmary. Up to this time the Infirmary, as a Teaching Hospital, was directly managed by the Department of Health but now it was the responsibility of the Area Health Authority and the Regional Health Authority. What made matters considerably worse was the fact that the senior managers at St James's were transplanted to run the new Regional Health Authority. Thus our "enemies" from across the town were now our political masters.

In 1973 the country was judged to be ungovernable and Heath went to the country and lost very narrowly with a vote of Labour 301 votes and the Conservative 297.

It was not long after this when the Regional Health Authority announced that it was going to cancel the building of Phase 1 (now the Clarendon Wing). It was an almighty disaster as gynaecology was now being performed in temporary buildings in Roundhay Hall cut off from the main hospital. Problems occurred in the hospital service in particular at the Maternity Hospital in Hyde Terrace which was becoming more and more obsolete as time went on. The staff were working in very difficult working conditions as many departments had been put on hold to await the new hospital.

There was considerable political upheaval and some very angry meetings in Faculty (the Senior Medical Committee of the Infirmary). Someone arranged for questions to be asked in the House of Commons and letters appeared in national newspapers. In all this Professor McNicol took a leadership role and for this was admonished by the University who pointed out that he was a member of University staff and the Infirmary was nothing to do with him!

The Government became aware of all this turmoil and the site was visited by David Owen the Minister of Health. He decided that this area of urban spoil could not be left in this state and a true British compromise was reached. A new wing would be built but it would lose the top floor. Thus the spirited uprising of the consultant staff saved the day.

Shortly before this change over to the new system of regional administration Arnold Tunstall, Secretary of the Board, attended the Committee of Physicians. He announced that the Department of Health had awarded the Infirmary a new consultant in medicine. He asked if any specialty had any interest but was met with a wall of silence. The workload in haematology had increased and there was need for an additional post to concentrate on the development of Clinical Haematology so I volunteered to take the post much to the satisfaction of my colleagues who did not want to see their private practice diluted.

The post was advertised and this led to the appointment of Dr James Anthony (Tony) Child who had just completed a post of Medical Research Council Research Fellow investigating sideroblastic anaemia, with Professor David Mollin at St Bartholomew's Hospital, London. I was happy that Tony had worked with Mollin and would have learnt to practice haematology the Hammersmith way.

On the first day he started Professor McNicol asked Tony to do a locum for Dr Rex Tattersall, a physician with a distinguished war record in the Far East in the war against Japan, who had died suddenly. As I was not consulted I was not particularly pleased that Tony had taken this role when we had need of his full time presence in the Department of Haematology. However Tony enjoyed being a General Physician which he often considered being of a higher status than a mere Consultant Haematologist. For some time afterwards Tony signed himself Consultant Physician and Haematologist. But he was wrong, the days of the General Physician were ending and a General Physician would never be appointed again at the Infirmary, in fact, Dr Tattersall's replacement was Tony Axon a Gastroenterologist.

As soon as George McNicol got his feet under the table he set out to prise haemophilia from St James's. He held a series of meeting with me and Dr L A D Tovey, Director of the Regional Blood Transfusion Service, who had been appointed relatively recently. Dr Tovey followed Dr Shone, a non-medical scientist, who had developed his reputation by inserting thermo couples into the rectum of rabbits to test for pyrogens in blood transfusion fluids. The reason for the appointment of a non-medical scientist is shrouded in mystery and a thick veil of secrecy covers the event. As rumour has it the first medical director, and other medics, took teams of donor assistants, who were all women, to the East Coast to collect blood and this meant they stayed over night. It is rumoured that the sex orgies took place at night and this went on until one donor assistant considered she was not getting her fair share and complained. It is said that the then director took up a post in general practice in Leeds and disappeared from view.

There was therefore, on Shone's retirement, a need to appoint another director. The Region wished to appoint another non-medical Director but this caused controversy at both a national level and amongst local consultants and so there was an impasse. It was appreciated however that Dr Derrick Tovey, a Consultant at Bradford, had considerable experience in blood transfusion when he trained at the Middlesex Hospital in London and he was approached and offered the Directorship much to everybody's satisfaction. Once the Region appreciated that Derrick was not a sex predator everything went well and certainly as far as I was concerned in haematology when we began to receive blood component therapy from the Regional Blood Transfusion Service it considerably enhanced clinical care. From my knowledge of Regional Directors at the time there were no particular sex related problems but quite a few of the directors had delusions of grandeur. I knew several directors who were driven about their local hospitals using a chauffeur driven Rolls Royce.

The meetings between George McNicol, Derrick Tovey, and myself to discuss the takeover of the St James's Haemophilia Centre went on for some time without a serious plan. George had already determined that the Metabolic Ward would be the Haemophilia Centre. I assumed that George would appoint a Senior Lecturer to run the haemophilia service but when he met with Layinka Swinburne to talk about the future of haemophilia in Leeds he told her that the Haemophilia Centre at the Infirmary would be run by Drs Roberts and Child. As Layinka Swinburne knew both Tony and I had little experience in managing haemophilia the negotiations came to an abrupt end.
George McNicol, Derrick Tovey, and I however continued to meet as we had been invited by the Royal College of Pathologists to undertake the practical examination for the final of the Membership in Haematology in Leeds.

I acted as principal organiser and was the main line of contact with the College. George McNichol was the leading figure in the clinical examinations which were held in the Littlewood Hall in the Infirmary and Dr Derrick Tovey organised the blood transfusion examination in the laboratories at the Regional Blood Transfusion Centre at Seacroft. I was responsible for organising the rest of the practical examination in coagulation, morphology, and various aspects of anaemia.

At first we had, until the Clarendon Wing was opened, no space so we had to hire laboratories in Leeds Metropolitan University which was just across the road. In the Haematology Department Geoffrey Tate who was the Chief Technician in Coagulation set up the examination in coagulation. This was based on the simple coagulation tests based on the activated partial thromboplastin time and the prothrombin time and the marks were allocated more for the interpretation of results rather than practical expertise. Philip Day, the Chief Technician in the blood counting laboratory who was responsible for other investigations of anaemia, such as interpreting the results of haemoglobin electrophoresis, usually prepared some test haemoglobin electrophoresis strips and those for glucose-6-phosphate dehydrogase deficiency. I personally looked after the morphology, all slides had to be beautifully spread, stained and mounted. At all costs I would avoid the horrors of getting my microscope dry lens coated in oil, which I had experienced when I took the MRCPath myself.

Charles Buchan the Senior Chief Technician acted as a general factotum and organised the candidates, telling them where to go in particular on the day of the viva when candidates were interrogated by two pairs of examiners. There was always an external examiner. When organising the exam I concentrated on simplicity with no Machiavellian tricks. As I knew to my own cost, the stress of the examination was enormous and the examination itself had to be straight forward but a true test of the ability of the candidate.

The stress was obvious and Charles Buchan often acted as a father confessor. I remember one girl in tears at the beginning of the exam as her husband had recently died, and in the middle of the practical examination one candidate collapsed due to a pneumothorax. He was admitted to the ward and that was the end of the examination for him. It meant however that we had to organise half a practical examination at a later date although we were allowed to continue this without an external examiner. On another occasion a Muslim candidate asked for a prayer room into which he disappeared at regular intervals throughout the examination.

We were heavily dependant on our technical staff throughout for organising the practical examination and for which they received no award except for the fact that there was an examiner's dinner in a nearby Chinese restaurant to which all were invited. Dinner was usually paid for by me.

At the end of the examination I usually thanked God that nothing had gone wrong and that there were no mix ups. On one occasion I had to act as external examiner at the Hammersmith Hospital, and all appeared to go well, in particular there was a candidate I had examined before who was known to have difficulty in passing the examination. On this occasion however all went well and we were quite happy to pass this candidate. After the results of the examination had been published I received a telephone call from the consultant this individual was working for. In desperation he said that the candidate had failed yet again and could I give any advice as to where his weaknesses were and what he should do to get through. To my horror I remembered that in actual fact he had passed the examination. I therefore asked the consultant to wait for a time and then I would ring him back. I rang Professor Lucio Luzzato at the Hammersmith with this quandary. He checked the records and found that he had succeeded as I thought. He immediately telephoned the College and they had to admit they had made an error and had wrongly allocated him a failure. If the candidate's consultant had not telephoned me nobody would have ever been the wiser! Another problem arose when the examination results were published in the College Bulletin and yet again they forgot to include the candidate's name. This led to a personal apology being printed in the next Bulletin from the President of the College.

As a Chief Technician, Geoffrey Tate, produced a superb service. He was knowledgeable, well read, and appeared to keep up with the literature. When any problem occurred in the laboratory from a screening test then Geoffrey would go ahead, visit the patient, take further samples, and by the time a member of the medical staff reached the Coagulation Laboratory he usually had the diagnosis ready. He was of great help to the Professorial Medical Unit. He looked after their Coagulation Laboratory, taught medical staff techniques, and kept an eye on the overall standards. This role in the Department of Medicine resulted in him working closely with Dr Andrew Davies, Senior Lecturer in Medicine. Under Andrew's supervision Geoffrey produced a PhD thesis on the hypercoagulable state in promyelocytic leukaemia and which led to him presenting a paper at the annual meeting of the British Society for Haematology.

Geoffrey's other interest was in the lupus anticoagulant. Systemic lupus erythematosis is an autoimmune disease with various systemic manifestations but it is often associated with this hitherto benign abnormality. It was found lupus anticoagulant was often found in patients with recurrent abortions. This interested the Professor of Obstetrics and Gynaecology, Professor James Scott, who enlisted Geoffrey into his study of the condition. Another participant in the study was John Turney, a Renal Physician who was an expert on renal dialysis, who performed plasmaphoresis on some of the patients. It was pleasing that the wife of one of our former registrars who had recurrent abortions was treated by plasmaphoresis and had a successful pregnancy.

Thus Geoffrey Tate was a major figure in the Department and was highly respected around the hospital. I was very pleased with Geoffrey's success and it was my policy that should any of the technical staff show similar promise then they got my maximum support and some time off to pursue their studies.

One benefit of working closely with the Professorial Medical Unit was the fact that they attracted some very good academic staff. In particular John Cawley came to work in the Department. He was a Leeds graduate and subsequently worked with Professor Frank Hayhoe in his department at Cambridge. I was a particular admirer of the department at Cambridge. Frank Hayhoe and his clinical colleagues ran a very active clinical department and ran the MRC Trials in Acute Myeloid Leukaemia. They had a very high reputation for leukaemia diagnosis and Frank Hayhoe and his Senior Chief Technician, Roger Flemans, had published a book on

Leukaemia Diagnosis and Histochemistry. John Cawley whilst working in these laboratories had developed an interest in electron microscopy and was an expert in this technique in the field of leukaemia. He came to Leeds knowing that there was an electron microscope in the Algernon Firth Institute which he began to use with the permission of Professor Lumsden. It was obvious from the outset that John and I had a good deal in common with our interests. In particular at this stage I was acting as a regional referee in matters of morphological diagnosis and I had quite a wealth of material. John Cawley had a good working relationship with a scientist called Gordon Burns in the Cambridge laboratories and thus we were able to set up several collaborative projects in this field. John had an overwhelming passion, which lasted for the rest of his career, for hairy cell leukaemia and published two monographs on this topic. My success in diagnosing hairy cell leukaemia had got me a job at the Hammersmith and so our passions were mutual. He was excited to find that most of the cases occurring in Yorkshire were referred to me for diagnosis and so had much material to work on. John had a consuming interest in research and would regularly get fresh material in Leeds and then drive down the A1 at great speed in a battered old Ford Cortina to the Cambridge laboratories. He would often forget as he did this, that in actual fact, he should have been working in Medical Out Patients and as a result was highly unpopular among his colleagues for the number of times he failed to turn up. However some interesting and significant research papers emanated from this co-operation which for me was highly satisfactory.

All that we could do in Leeds to provide leukaemia diagnosis was very basic histochemistry using the peroxidase test to diagnose acute myeloid leukaemia and the PAS test in lymphoblastic leukaemia which was an unreliable marker for this condition. Such was the space problem in the haematology laboratory that these were very often performed on a window sill.

John was a very interesting character. His father was a Professor of Medieval English at Leeds University and was an expert in the Wakefield Mystery plays. His mother was a distinguished children's author. John therefore had a great deal to live up to and would have considered himself a failure in life if he did not achieve a Chair. A crisis appeared however when John was given the job of lecturing the dental students in haematology. The situation then, as now, was that the dental students were expected to learn all about haematology and its many ramifications in a one hour lecture. This was obviously stupid and unsatisfactory but it persisted and despite letters from me to the Dean of the Dental Institute nothing ever changed. John began his lecture by saying to the students that this task of teaching all of haematology in one hour was impossible therefore would they ask him questions and he would do his best to answer them. The students complained to the Dean who passed these complaints onto Professor McNichol who sent for John Cawley, went berserk, and tore him to shreds.

This was too much for John and before long he left Leeds and went to work in University College Hospital, London, where Professor Ernest Huens, a distinguished haematologist with an outstanding research record, was in charge. From there John was successful in obtaining the Chair in Haematology at Liverpool University.

At the same time, as a Registrar we had a young doctor, Sweelay Chein, who was obviously a very gifted young doctor. She left and obtained a post at the Molecular Medicine Unit in Oxford University with Professor David Wetherall the world leader in DNA studies of haemoglobinopathies. It appeared to me that my little world in haematology was beginning to disintegrate. We had lost Sweelay Chein and John Cawley. There was not a square inch of space found in the Department and nothing would change until Phase 1 was built.

However about this time a young Dr Milligan was appointed as a Lecturer in the University Department of Medicine. His main interest was in haematology and he wished to pursue some research. His research interests lay in the field of bone marrow transplantation and he wished to do some animal work. Professor McNicol referred him to Professor Colin Bird in the Algernon Firth Institute to discuss a research project. This ended in Don Milligan and Professor Bird mutually regarding each other as idiots and Dr Milligan returned to Professor McNicol destitute of ideas. Professor McNicol asked to see me and asked me for my help with Dr Milligan. At that time there was currently a good deal of interest in blood flow in polycythaemia vera, particular the patients with iron deficient red cell indices. I also knew that the other major interest in the Department of Medicine, apart from haemostasis and thrombosis, was in diabetes which had continued since Professor Tunbridge retired. As part of these studies there were pieces of equipment for measuring and visualising blood flow in capillaries. I suggested to Professor McNicol that it would be useful to apply the technology available for vascular studies in diabetes to polycythaemia. This was put to Dr Milligan who grudgingly agreed to do the work but ended up doing it extremely well with several good publications. He used the machine which made a visual record of blood flow in capillaries to compare the sluggish blood flow in chronic myeloid leukaemia with a high granulocyte count with that of normal flow in chronic lymphocytic leukaemia with a high lymphocyte count. This was presented at the 1983 meeting of the British Society of Haematology and was very well received. Following this Don got a rotational senior registrar post in haematology in Yorkshire and finally moved to be a Consultant Haematologist at Birmingham.

In 1980 George saw me and told me that he had been invited to be the President of the British Society for Haematology in 1983. The meeting would obviously be held at Leeds University and he asked me to be the Meeting Secretary. I was very pleased indeed to accept this, presenting the meeting would be a challenge but it would put haematology in Leeds on the map, and this excited me greatly. Two months later however he saw me and told me that he was leaving his post as Professor of Medicine in Leeds to become the Vice Chancellor of the University of Aberdeen. He went on to tell me that this was a life long ambition and that when he was a small boy when asked, what he would like to do when he grew up, instead of replying airline pilot, policeman, or fireman, he would reply, even at a very tender age, that he wished to become a University Vice Chancellor.

Thus George went off to Aberdeen to realise his life's ambition and when he got there he nailed the blue flag of Mrs Thatcher to the University mast. The rest of the story will be left to the reader's imagination.

This move of George's considerably upset the British Society for Haematology plans and they asked me if I would continue to be Meeting Secretary and organise the meeting in Leeds. They asked Dr Mitchell Lewis, Reader in Haematology, from the Hammersmith Hospital to be President for that year and he accepted.

This was very welcome to me because I knew Mitch Lewis very well from my time at the Hammersmith and from Scientific Meetings ever since. In the organisation of the meeting I enlisted the help of Charles Buchan, the Senior Chief in Haematology, and Ken Major, Senior Chief Technician in Blood Transfusion. Both knew how to organise a project, they were very practical and I was very pleased to have their help. The meeting began with a reception at Temple Newsome Hall, a Jacobean mansion on the outskirts of Leeds, which is owned by Leeds City Council; there was a buffet meal at the Stables Court Yard café followed by a tour of the mansion which contained a magnificent collection of antique furniture including several items by Chippendale and some outstanding works of art. The next day there would be a dinner at the Civic Hall and this would be preceded by a sherry reception in the Civic Art Gallery which contains many art treasures. As for the meeting itself, I for the first time organised a formal Trade Show. I also, for the academic program, organised individual sessions for different disciplines.

The Trade Show went very well and made some money but there were complaints about the specialist sessions as some contributors wished to speak to the entire Society in a plenary session. These difficulties would be resolved and this would be the pattern of the meeting for years to come. The meeting was a success and as a result I was elected to the Committee of the Society and from there I went on to be Associate Secretary and Secretary of the British Committee for Standards Haematology, then the Secretary of the Society itself when I, with colleagues, undertook a major

141

reorganisation of the Society and made peace with Blackwell Publishers.

There had been a considerable amount of enmity between the Society and Blackwells because Blackwells owned the British Journal of Haematology which anteceded the Society. Blackwells wanted all members to receive the British Journal of Haematology but senior members of the Society would not agree. This eventually was resolved much to the great financial benefit of the Society. Finally I was elected President of the Society in 1993-94 but this is not the major theme of this book.

The Professor of Cancer Research

Professor E H (Teddy) Cooper was Professor of Cancer Research. He came to Leeds in the late 1960's following a rather traumatic and unfortunate end to the old Department. Professor Passey, a distinguished figure in the world of cancer research unfortunately hit the bottle and I can remember very well seeing him staggering along Great George Street in a highly inebriated manner. In the late days of his Department there were all sorts of problems, including financial ones, and as a result most of the department was shut down and by the time Teddy Cooper took over the Department was restricted to the Professor and about three members of staff in a few rooms in the Worsley Building.

Teddy was well known for proselytising young bright men. He brought with him to Leeds Dr Sunitha Wickramasingh who had done some work in Cambridge with Cooper and Chalmers, the Consultant Haematologist in Cambridge at the time, on the cell cycle by labelling cells with tritiated thymidine. At the time there was a day symposium held by the British Society for Haematology on the cell cycle with lectures from Cooper and Chalmers and I asked Sunitha why he had not attended this meeting. He told me that there was no point in him going to hear other people present his work. He was obviously a bright young man but did not stay long in Leeds and went back to London and eventually became Professor of Haematology at St Mary's Hospital.

Teddy Cooper therefore began to form an association with me and told me that he had now dropped any consideration of the cell cycle and was now going to work on tumour markers. I was not clear what a tumour marker was but what Teddy proposed to do was measure a series of biochemical estimations in a whole range of biological substances, most of which were commonly used as tests in chemical pathology. I must admit that in haematology we were well suited for this co-operation because in our pre-clinic meetings we were able to organise any tests we wished on any patient and so that all we needed to say was "blood for Teddy Cooper" and an appropriate sample would be collected, labelled, and sent to him. The first substance he chose was gamma glutamyl transpeptidase a measurement of which is commonly used as a liver function test. He investigated lymphomas and leukaemias by performing this estimation. Teddy then presented me with the results and asked me to present them at a meeting of the Association of Clinical Pathologists in York.

I attended the meeting and found that I was second on the programme. The first presentation on the programme was to be by Dr Chalmers of York, a known eccentric. The lecture theatre was full, and at the front of the lecture theatre there were two projection screens, one to the left and one to the right. Dr Chalmers began by looking up at the left hand projection screen but the projector sent the image completely the other way onto the opposite screen at which Dr Chalmers had to run across the room and all the audience turned their necks. This was but the beginning. As he went on he became more dishevelled, waved his arms, his hair began to stand on end, and all this was received with a good deal of laughter. Dr Chalmers had a colleague in haematology, Dr Cedric Wiley, who sat at the back and made innumerable cat calls and noises like a cock crowing which all added to the air of confusion. Dr Chalmers ran his department in a very idiosyncratic way. One thing he did with a Coulter automated blood cell counter was to take the printed results form and instead of sending this out to the ward the results were copied by hand onto cards and stored on a vast rotating drum in the cellars of the hospital in the middle of York. This was very tedious for the technicians but they found their salvation when the River Ouse flooded the cellar of the hospital and destroyed all records.

I remember again on one occasion going to a Blood Club meeting where Dr Chalmers presented a case. It was the custom to produce post mortem findings on photographs projected onto a screen but on this occasion we were surprised to see Dr Chalmers stagger into the lecture theatre with a cadaver which was in rigor mortis but which had had a post mortem performed. He then proceeded to pull back the abdomen, take out organs which had been fixed in formalin, and demonstrate them. When this was completed he again staggered out of the room, down the corridor, and back to the post mortem room. This was almost unheard of behaviour.

Cedric Wiley was another eccentric. It is said that when he was a student he trained as a British Rail guard and worked in Scotland on his vacation. Apparently there was a major problem when the train went to pick up passengers at Gleneagles, a luxury hotel with a championship golf course. There was considerable concern among the porters at the station in that the luggage from wealthy customers was brought down to the station by porters from the hotel who then carried the luggage onto the train and received the tips. Thus the porters on the railway station received nothing whatsoever. So what Cedric did was to wait until all the porters from the hotel were on the train and talking to the guests of the hotel then blow the whistle to start the train, all the doors were locked, and the hotel porters were taken on to the next station from whence they had to find their own way back.

Thus the presentation by Dr Chalmers continued with considerable uproar until the end when, in a highly amused state, virtually all the audience disappeared and I presented my paper, the next one, to an audience of about six.

Fortunately for me when Tony Child was appointed Consultant Haematologist he formed a very useful and profitable co-operation with Teddy. Pharmacia, a pharmaceutical company, produced a test to measure a substance known as Beta-2 microglobulin and asked Teddy Cooper to measure the levels of this in several disorders. Derek Norfolk was at the time the Senior Registrar and the substantial data was presented to him to sort out. He analysed the data and found out that the levels of Beta-2 microglobulin were highly significant in the stratification or staging of myeloma. This was a highly significant piece of work and was adopted world wide. Teddy Cooper and Tony Child traversed the world presenting their results and without doubt they established Leeds as a Centre for the study of myeloma.

Teddy Cooper never ever again found a substance of similar significance in his work on tumour markers, but he began to have a political influence in the Yorkshire region. Teddy and Charles Joslin, the Professor of Radiotherapy at Cookridge, worked together and organised and promoted an organisation known as The Yorkshire Regional Cancer Organisation. This organisation was to promote interest and knowledge of cancer within the Yorkshire region, to co-ordinate workers in the field, and to promote the institution of clinical trials. At Cookridge Hospital several portocabins were erected and staffed by workers in the field and several secretaries. This was invaluable to me, as the Yorkshire representative on the Medical Research Council Committee, for therapeutic trials in leukaemia, as I was obliged to organise regional meetings of consultants to present results and to encourage the consultants to enter patients into the MRC trials.

The Pathology Professors
In my first few weeks as a Consultant Haematologist and Head of the Department of Haematology I realised there was precious little contact with other Heads of pathology departments and Professor Lathe, Head of the Clinical Pathology Department, I had yet to meet to have a chat. The Head of the Department of Microbiology was Professor C L Oakley, FRS. Microbiology was a major Scientific Department of the University and the hospital service was but a part of this. The Head of the Clinical Service was Dr Kurt Zinnenan, Reader in Microbiology. I had got to know him because he spent a considerable time in the Infirmary and was a useful friend. It was obvious that Kurt did a little private practice on the side for one day he asked me to perform a blood count on a private patient and

arranged for the patient to come along to the Department. I was surprised to see a gleaming Rolls Royce turn up outside the Haematology Department among the ambulances and for a multi-millionaire from Bradford to descend and enter the Department. The blood count was normal but I was expected to charge a fee.

The request to perform blood counts on patients by general practitioners as a domiciliary visit began to increase. I was not very keen on this as to go find addresses among the streets of Leeds and perform a venepuncture could take the best part of half a day but yet I continued to receive these calls. Dr Eric Allibone, a retired paediatrician from the Infirmary who had taught me as a student, had retired but taken up general practice. I got a telephone call to go and visit an elderly lady for a blood count and I felt I could not refuse him. I went, did a blood count, diagnosed iron deficiency and left it at that. However Dr Allibone called me back to see how the patient was responding to iron therapy, so along I went. Dr Allibone was waiting with the patient for me and said to the patient "go on, give him the brown envelope". She handed me a brown envelope which I placed in my pocket. She then said "there's £5 in there but I don't see why I shouldn't get a blood count free from the NHS". I was somewhat nonplussed, but then the patient said to me that she did not believe in what she had just done, but Allibone had made her, and anyway she was going to complain to her brother who was a member of the Regional Health Authority. I thus foresaw that my career in haematology as a National Health Consultant could now become a record breaking shortest appointment as a consultant ever.

However things blew over but I was unhappy doing all these domiciliary calls even though they were well paid. I was called by another general practitioner to perform a count. I left a message with his receptionist that if he wanted a blood count he could do it himself. This led to an official complaint with the District Health Authority. However my mind was made up, I was not doing any more domiciliary visits as this would take up the best part of my time when I had a lot of work to do to develop the Department. I therefore decided to provide an alternative general practitioner service. I, with Philip Day, the Chief Technician in charge of the routine blood counting laboratory, designed a pink request card for

146

haematology investigations. These were distributed to the general practitioner who had the alternative of having the blood taken in his surgery and sending it down or, in fact, sending the patient down to our phlebotomy suite which was handily placed in the Brotherton Wing opposite the Civic Hall. This worked well and the phlebotomists took this extra work in their stride.

Things however were changing in the world of Pathology. Professor C L Oakley had retired and was replaced by Professor Douglas Watson who was an internationally acclaimed expert on the herpes virus. A new professor, Professor Mary Cooke, was appointed to run the Clinical Service to the Infirmary and before too long Professor Lumsden collapsed and died from a myocardial infarction. There was therefore an interregnum before a new Professor of Pathology was appointed.

The Department of Health now introduced a new initiative to promote medical management in the NHS. This was called the "cog wheel system". Each discipline, such as, medicine or surgery, or pathology, would form a Division and each Division would elect a Head, or Chairman, and these would report to a central committee known as the Executive Committee which comprised all chairmen of divisions. At this time the Medical Advisory Committee comprised an organisation known as the Faculty. This dated back to the time before the University of Leeds was created in the early part of the 20th century. The early consultants at the Infirmary before the University was formed, all taught medical students and when they met monthly to discuss Infirmary business they called themselves "The Faculty". This Faculty continued with this function for many years. At first the number of members were small and they would meet in the Board room for a pleasant meal and then after the port and cigars would discuss the few items that were to be resolved. The Faculty however grew in size as more and more consultants were appointed, but when I was a consultant the Faculty was still in full flow. It had a lot of members but would meet once a month to discuss business often put to it by the management to decide collectively on what action should be taken. The privilege of sitting

147

on the Faculty was regarded very highly and when the new "cog wheel system" was put to them they rejected it outright.

When the University was created they set up a Faculty comprising the University teachers but the Infirmary staff retained the title Faculty for their monthly meeting.

At this time the University Departments of Pathology was paid a university budget for teaching, research, and for routine service to the hospital. Each year there would be a negotiation between the University and National Health Service and a sum of money would be agreed for the Infirmary to pay the University for the service it provided. Thus there was no need for any hospital advisory scheme. However the new "cog wheel system" had various functions such as the allocation of finance for equipment. I therefore, in Haematology, needed a cog-wheel system to manage my relationship with the hospital management. I went to speak with Richard Oswald, the District Administrator, to discuss my problem and my relations with the other University Pathology departments. After consultations with Department Heads it was proposed that there should be a scientific cog wheel function and this would comprise the Divisions of Pathology, Radiology, and Medical Physics. A Scientific Services Executive would comprise divisional chairman and secretaries. This was put to colleagues and they were all willing to participate.

Professor Grant Lathe, Head of Department of Chemical Pathology was elected Chairman of the Scientific Service Executive, and George Reed from the Department of Medical Physics, the Secretary. It was agreed that the divisions should be composed of the Heads of Department and the Senior Chief MLSO's. One of the Department Heads would be elected Chairman. In Pathology it was appreciated that the same representation should apply to the Hospital for Women and the Maternity Hospital.

Once a year each Department had to put forward requests for items of equipment, so each Department had to fill in a request form, and make a list in departmental priority order. It was then collected together at Division and rearranged in a Division of Pathology Order and then taken to the Scientific Services Executive Committee and they would be again drawn up into a Scientific Services Executive

Committee Priority Order and sent to the District Health Authority for funding. The District Administrator would decide how much money was available in each category. In theory the scientific list should again be arranged in priority order with the medical list but this did not take place as the Faculty refused to co-operate. This meant that for some years the Scientific Service Departments were very well funded and our Scientific Divisional system began to work very well.

In 1975 Professor Colin Bird was appointed Professor of Pathology. Colin came from Edinburgh and was the third Professor of Pathology from Scotland that century. In the 1920's Matthew Stewart was appointed Professor of Pathology. He was followed by Professor Rupert Willis, an Australian, who in turn was succeeded by Professor Lumsden an Aberdonian Scot. So therefore Professor Bird continued this tradition and until the turn of that century no English professor had ever been Professor of Pathology. Colin was a typical Scottish professor who believed the Head of Department should wield absolute power.

He often said that you ask an Oxbridge student "what do you think", a red brick university student "what are you taught", and a Scottish student "what does the professor tell you". There have been cultural links with France and Scotland since the time of Mary Queen of Scots and Scottish universities seem to follow very much a continental pattern of administration of departments. However Colin was an extremely pleasant chap and very easy to get on with and he had one great overwhelming virtue, his main interest was in the lymphomas, so this complimented the Department of Haematology in a very big way. We would in future co-operate in research projects and there would always be an expert in histopathology of lymphomas and this would be an enormous help with the lymphoma trials.

Thus Pathology was organised to face the future and the many developments and challenges it would meet in the forthcoming years.

CHAPTER 9

HAEMATOLOGY IN THE LEEDS EASTERN DISTRICT

"East is east and west is west and the twain shall never meet"
The ballad of East and West by Rudyard Kipling

But as far as Haematology was concerned the introduction of high technology meant they had to!

In the post war years pathology developed along two different lines in Leeds.

The Infirmary was served by specialist university departments each with an eminent professor as Head. On the other hand at St James's the pattern of pathology was that of clinical pathology. These laboratories were staffed by multi disciplinary pathologists who were expert in all fields of pathology. In fact these medical clinical pathologists spent most of their time practising histopathology and performing autopsies. The individual divisions in pathology had a senior chief, or chief technician, running them and who would seek help from the consultant if a diagnostic problem occurred. In Haematology the consultant would perform bone marrow aspirations and report them but in the case of St James's, Dr Bill Goldie the Head of Pathology, also set up and was Head of the Regional Haemophilia Centre.

As far as Haematology was concerned at St James's the Senior Chief in charge of haematology was Roger Hall a highly intelligent young man who ran an excellent department, taught at the Polytech, and eventually wrote a text book on Practical Haematology which rivalled that of Dacie & Lewis. He collaborated with Dr Monty Losowsky who, from his early days in the University Department of Medicine, had an interest in haematology. Monty Losowsky went on to become Professor of Medicine at St James's. In September 1963 a patient was admitted to St James's under the care of Professor Losowsky with a severe haemorrhagic disorder and Roger Hall set out to investigate him. However all the coagulation tests were normal and it was decided that in view of the severity of this disorder to send samples to Dr Rosemary Biggs at the renowned Oxford Haemophilia Centre for her opinion. She rang back to say that this was a case of fibrin stabilising factor deficiency and only two to three cases had been described in the world. Losowsky and Hall went on to find five more unrelated cases in Yorkshire. The patient was treated with fresh frozen plasma successfully. Fibrin stabilising factor is the thirteenth of the coagulation factors which are important for production of a blood clot in the laboratory, or thrombus within the living circulation. It is now available as a concentrate and can be used for maintenance therapy.

Studies on the conversion of fibrinogen to fibrin has been a Leeds interest for many years. As mentioned in Chapter 1 Charles Turner Thackrah studied this reaction calling the clot the crassamentum. Thackrah determined the normal range for fibrinogen and showed how the clotting time varied in relation to several medical disorders. After the discovery of X-rays by Roentgen in 1895 there were investigations as to the nature of these rays and whether or not they behaved like light. Von Laue investigated them to see if the atoms of a crystal would act as a diffraction grating. With colleagues he found that a crystal of copper sulphate would form and diffract x-rays and produce bright spots on a photographic plate. This aroused the interest of Lawrence Bragg who saw that this technique could provide information about the way in which atoms are arranged in a crystal. Lawrence Bragg was born in Adelaide in 1890 where his father, William Henry Bragg, was Professor of Mathematics and Physics. In 1909 W H Bragg was appointed Professor of Physics in Leeds and Lawrence went to Cambridge to study Physics at the Cavendish laboratories. When Lawrence Bragg heard of the German work he was greatly interested and worked out a simple mathematical rule, now known as Bragg's Law, that opened up the possibility of using crystallography to find the structure of molecules. The Braggs, then in Leeds, went on to show that in crystals of sodium chloride the crystals were made up of a regular lattice of alternating sodium and chlorine atoms. They then went on to show how the carbon atoms in diamonds are arranged. In 1915 while Lawrence was serving in the trenches he heard that he and his father had won the 1915 Nobel prize for Physics.

In 1923 W H Bragg, the father, was appointed Director of the Royal Institution in London where he continued work on X-ray crystallography. He had two bright young men work for him: Desmond Bernal who was interested in extending X-ray crystallography into organic compounds, and William Astbury who was interested in similar work in organic fibres and managed to get an X-ray pattern from a pattern of alpha-keratin. Subsequently Bernal went to work in the Physics Department in Cambridge, and Astbury came to the Textile Physics Department in Leeds University. Bernal went on to work on proteins, in particular the enzyme pepsin. Soon there would be some significant movement in staff. Lawrence Bragg left the Chair of Physics in Manchester to be replaced by Blackett who moved from Birkbeck College in London. This vacancy was filled by Bernal. Lawrence Bragg went to Cambridge to be the Cavendish Professor of Physics in 1938.

A promising young student at Cambridge was Dr Max Perutz who was very interested in the molecular structure of haemoglobin (the chemical structure was already worked out and was 4 porphyrin rings each with an attached iron atom and two globin chains, an alpha chain, and a beta chain) and greatly impressed Bragg with his work. He was also joined by Kendrew who began to work on the structure of myoglobin. In 1947 Bragg approach the Medical Research Council asking for an MRC grant to cover the work of Perutz and Kendrew and assistants which was successful. This went on to become one of the most successful research units in the world.

152

It was joined by Crick and Watson who, in 1953, published the double helical structure of DNA.

Lawrence Bragg was already a Nobel Laureate but Perutz, Crick, and Watson went on to join him.

Thus the pioneering work on X-ray crystallography in Leeds in the early 20th century developed through the Royal Institution into two arms, one at Cambridge which produced one of the best research laboratories in the world, and the other at Leeds.

In 1928 William Astbury, FRS, was appointed to the University of Leeds Department of Textile Industries. Astbury had a particular interest in wool which had the ability to stretch and contract. He performed studies with his student, Alex Street, and they were able to take x-ray diffraction pictures of keratin in a natural folded alpha state and stretched beta state. In 1938 Astbury also did x-ray diffraction studies of DNA when no one knew what the function of DNA was. He described the structure of DNA as helical. This was incorrect but ultimately led to the double helix structure described by Crick & Watson in 1953. Astbury was responsible for introducing the term molecular biology. Current research in Astbury's department today is looking at the transition of fibrous proteins in alpha and beta forms into amyloid protein in the brain in both Alzheimer's and Parkinson's diseases. In the late 1940's a young Hungarian student by the name of Lazlo Lorand came to work with Astbury. He used the fibrinogen-fibrin model for his work and was able to describe several types of blood clot and postulated the existence of a fibrin stabilising factor which was highly relevant to the studies at St James's. Monty Losowsky contacted Lorand and subsequently they became collaborators and friends. Monty Losowsky has continued this interest in fibrin stabilising factor, now Factor XIII, throughout his career. Now Professor Peter Grant in the University Department of Medicine has further continued this work in this field by investigating the phenomenon of myocardial fibrosis in Factor XIII deficiency and attempting to assess its role in this disorder.

Monty Losowsky's other major interest in haematology was that of congenital sideroblastic anaemia. In his early days he toured Great Britain finding members of a kindred with this disorder and published his findings. Throughout this he was assisted again by Roger Hall who did all the morphological examinations of the film and bone marrow and made the laboratory diagnosis.

In 1976 the Standing and Advisory Committee in Haematology of the Joint Committee on Higher Medical Education visited Leeds to assess training in haematology. This was led by Professor Peter Flute from King's College Hospital, London and training was deemed unsatisfactory at St James's because of the lack of a specialist consultant haematologist and recognition for training was withdrawn until such time as a consultant was appointed. This did not go down very well at St James's Hospital as all the pathologists were confirmed clinical pathologists and did not want this pattern disturbing but there was no alternative and eventually they had to accept this decision with poor grace. So a new consultant haematologist was approved and finally Dr David Barnard, a Senior Registrar in Edinburgh, was appointed Consultant Haematologist at St James's in February 1977. David therefore came into a job where in fact he was not really wanted. The laboratories were under the control of the Head of Department, Peter Dosset, and to cap it all there were no beds allocated to the post which was a disgrace. The fault probably was with the Regional Advisor who assessed the job description and should have blocked it without guarantee of beds. This was yet another example of the difficulty in establishing the new specialty of haematology in Leeds and Yorkshire.

David Barnard was however a pleasant chap and very good at his job and was soon accepted by all.

There was plenty to do as no one in particular, except for the radiotherapists, were treating malignant haematology at St James's.

Also at this time paediatric oncology moved to St James's under the leadership of Dr Clifford Bailey.

During the late 1970's the technique of bone marrow transplantation had been introduced into Great Britain and the hospital that was pioneering this technique was the Royal Marsden Hospital in London. Bone marrow transplantation was obviously, despite its problems, a very successful form of treatment and patients were now living who would have otherwise died. Many hospitals throughout the country were considering introducing bone marrow transplantation and so the National Health Service appointed Sir Douglas Black, a highly respected former President of the Royal College of Physicians to make recommendations about the number and distribution and siting of bone marrow transplant centres. At the Infirmary it was decided to wait for the Black Report before introducing bone marrow transplantation.

One day John Cawley was visiting David Barnard to discuss a case and while David was out of his room John Cawley could not but help notice documents all over his desk, prepared with the co-operation of Cliff Bailey, for a submission to the Regional Health Authority for recognition of St James's as a bone marrow transplant centre. Word obviously got back to the Infirmary and Dr Tony Child decided that we at the Infirmary should also make a submission for recognition as a bone marrow transplant centre. This led in time to an "across town" working party including the radiotherapists at Cookridge. It was finally agreed that allogenaic bone marrow transplantation ie bone marrow from a donor, be carried out at St James's Hospital and that mainly autologous bone marrow transplants ie bone marrow withdrawn from the patient and after treatment given back to the patient, be carried out at the Infirmary.

In 1981 the Regional Health Authority accepted this proposal and agreed to fund it. There would be adult paediatric transplant beds at St James's and funds for radiotherapy at Cookridge. At the Infirmary it was agreed to use a four bedded sided ward on Ward 29 for the treatment of patients with acute leukaemia which would comprise an upgrade of the total side-ward and the provision of a sterile en-suite cubicle. Money was available for extra nurses. The Friends of the Leukaemia Unit took the lead in the plans for the four bedded unit, including the cubicle, and contributed £35,000 to the design and construction of this ward. *(Chapter 10)*.

The first allograft took place in 1982 in a paediatric patient. Dr Derek Norfolk, then a Senior Registrar in Haematology at St James's, was sent to the Royal Marsden Hospital for several weeks to learn the technique of bone marrow transplantation and when he returned he performed the first allogenaic bone marrow transplant in an adult in Leeds.

The practice of performing bone marrow transplants continued to increase and soon began to outgrow facilities. This would be finally resolved with the opening of the Bexley Wing at St James's where both units were merged.

When Dr Tovey was appointed in the late 1960's the Regional Blood Transfusion Service was very much run down and took a long time to crawl up to modern standards. I remember very well Dr Crossland who was confined to a wheelchair because of poliomyelitis in the past did however a tremendous job providing a very expert service in serology and that he was responsible for identifying most of the abnormal antibodies that occur as a result of blood transfusion.

The appointment of Dr Tovey followed a break through in the treatment of haemolytic disease of the new born. Sir Cyril Clark in Liverpool, who was a geneticist and very much interested in genetics of butterflies, transferred his interest to the Rh group of antigens. He became aware of haemolytic disease of the new born and sought to find a solution to the problem. Cyril Clark immunised male volunteers with RhD positive cells and raised antibodies against this antigen. When a woman was suspected of having haemolytic disease of the new born foetal red cells were sought in maternal blood by means of a Kleihauer test which is a blood film stained to reveal red cells containing foetal haemoglobin. If the test was positive they

156

were then given Anti-D which had been raised in the male donors and this successfully prevented the signs of haemoglobin disease of the new born developing. The Yorkshire Regional Blood Transfusion Service was unusual in that it did all the ante natal screening for blood groups and antibodies within the Yorkshire region. Dr David Tovey was able to collect data for all pregnant women in Yorkshire and amass a very large series of cases. This was David Tovey's major interest throughout his career. There is no doubt that if Clark's work was proposed today the local ethnical committees would not permit it.

As well as being Director of the Blood Transfusion Service, he had to act as Consultant Haematologist at Seacroft Hospital in whose grounds the Regional Blood Transfusion Service was situated. The Regional Paediatric Oncology Service was based at Seacroft at first and Dr Tovey had to provide a diagnostic service for patients admitted with leukaemia. He was also responsible for the haematology service at Killingbeck Hospital, formerly a small hospital but now a hospital for chest diseases and home to the Regional Cardiology Surgical Unit. When Dr Rajah was appointed as a second consultant he took over the haematology service at Killingbeck and actually acquired beds there. The situation was therefore that Dr Tovey and Dr Rajah as well as having major responsibility in the Regional Blood Transfusion Service were fully fledged consultants within the Leeds Eastern district and members of their Medical Advisory Service.

The third consultant to be appointed was Angela Robinson, very much a multi talented young woman. She was very accomplished in sport. She represented the county of Bedfordshire at hockey and tennis as a junior and represented St Mary's Hospital at squash at which she excelled. Angela, who was a very good looking woman, medical student, and doctor at St Mary's Hospital had techniques she developed for self preservation in the view of the large number of male medical students, Welsh rugby players, and doctors who inhabited St Mary's. Young men who took a fancy to Angela would often ask her for a game of squash. She would usually accept and meet her competitors in the courts. The young men were usually dressed in suitably tight and short trunks to exhibit their manly figure and they would then begin to have a warm up knockabout with Angela. She would then suggest that they begin to play a game of squash and would they mind if they used a match ball as Angela knew it would be very much more difficult to play with this. Such was her skill she would then proceed to completely out play the young man until he descended to the floor as a pile of sweat ridden, impotent, male flesh. Angela would then depart the courts with not a bead of sweat upon her beautiful brow and not one hair of her lovely blond hair out of place. Angela was and still is an accomplished singer of madrigals and she currently sings with the Yorkshire Chamber Music Choir. Angela came to Leeds and had joined the Regional Pathology Training Scheme as a Senior House Officer. The training in those days was devised to train clinical pathologists and therefore the first year she rotated to spend three months in haematology, chemical pathology, microbiology and histopathology. After a further year she took her primary MRCPath in haematology and chemical pathology and successfully passed the examination. A new senior registrar post, which combined haematology and blood transfusion, had been created and she successfully applied and obtained this post. She finally completed her training and was successful in obtaining the final MRCPath which caused Derek Tovey to become extremely agitated, for he was so keen to appoint her as consultant. Angela's husband, Joe Robinson, was a very good and promising radiologist and was looking for a consultant post. Derek was very concerned lest Joe got a consultant post outside Leeds and take Angela with him. Then Joe Robinson got a post in Leeds and Angela filled a newly created post in Blood Transfusion.

Haematologists in Leeds were all very keen to improve the blood product service and we were well aware that there was no cell separator in Leeds. It had been decided that there was insufficient work at either St James's or the Infirmary to fully staff and maintain and run a cell separator. Angela's first act, therefore, as a consultant was to acquire a Haemonetic 30 cell separator. She immediately improved the supply of blood products to the Region, but also pioneered the treatment of severe haemolytic disease of the new born by plasmaphoresis to remove antibodies. Angela's team of nurses began to use the machine for the routine production of plasma. Angela then set up at Bradford a Donor Plasma Donation Service using cell separators and this was the first in the world using voluntary donors.

In 1988 she became Director of the Yorkshire Regional Blood Transfusion Centre which immediately increased her administrative work so that she was no longer able to put in as much research work in the cell separator field. She had already acquired a considerable reputation within the United Kingdom for her pioneering work in the field of cell separators. A little later a review of the Blood Transfusion Service within England concluded that this should be run directly from London as a National Blood Transfusion Service and Harold Gunsen was appointed the first National Director. Angela shortly followed him in this position and thus became the Director of the English National Blood Transfusion Service.

We were all very proud of her success in Leeds. Not long afterwards Ian Franklin, a Leeds Haematology product, followed as Director of the Scottish Blood Transfusion Service which made me feel very proud of them both and our haematology training scheme in Leeds.

CHAPTER 10

PASTURES NEW

"At last he rose, and twitched his mantle blue: to-morrow to fresh
woods and pastures new"
John Milton - Lycidas

This chapter and the next describe the many changes that took place
in the NHS and the Medical School in the 1980's and 1990's. Many
were reactions to events both national and local and to changes in
personnel; only a minority were related to long term planning. If this
chapter appears incoherent at times it will be a true reaction to the
many events taking place.

In February 1978 the District Management Team in the Leeds
Western District issued a paper proposing the re-organisation of
Pathology Services in the Leeds Western District, including High
Royds, Wharfedale, and Cookridge. This began a process of visits
and discussions to produce some firm final proposals for pathology.

Of immediate concern were the hospitals at High Royds,
Wharfedale, and Cookridge. The four Heads of Department decided
to visit each site and meet with clinicians there as part of a fact
finding mission for, in all truth, the Heads of Department knew
practically nothing about what went on in these adjacent hospitals.
The first hospital we visited was High Royds, a large mental
institution. The four of us went up to a very large imposing door and
rang the bell. We were admitted and shown into a room and much to
our concern the door was pulled to and locked, so we were prisoners
in High Royds Hospital. We were subsequently taken out to visit the
laboratories and post mortem room and were amazed to be

transposed back into the late Victorian era. Techniques measuring haemoglobin were by methods I had not heard of and certainly did not appear in the standard text book of Dacie & Lewis. At the same time Professor Mary Cooke went to look at the hospital wards to assess the infection risks. She was amazed to find that before going to bed at night all the patients took off their socks and threw them into a large basket in the middle of room. When they woke up next morning they went to the basket and took any pair of socks that fitted them. We went to see an MRC Unit for the assay of psychotropic drugs which was in the charge of a Dr R P Hullin, who taught me biochemistry as a student. We then met with about twenty consultants, who were not pleased to see us, and their overall attitude was hostile.

We went on to recommend that the laboratory services, apart from Hullin's laboratory, should be transferred to Wharfedale, situated just down the road. The problem of post mortems at High Royds Hospital was discussed further with Colin Bird. As this would be a problem at Cookridge Hospital there needed to be a District policy for the performance of autopsies.

We then visited Wharfedale Hospital and met with the consultants, Dr Bates, Consultant Histopathologist, and Dr Antonis, Consultant Haematologist. We visited the laboratories and I in particular looked at the Blood Transfusion laboratory. The problem with Wharfedale is that it is situated some 15 miles from Leeds and the roads are virtually impassable at the early morning and evening rush hours. It would be impossible to produce a blood transfusion service, especially for emergencies, from the laboratories at the Infirmary.

A further visit to Cookridge Hospital followed which were under the consultant supervision of Dr Layinka Swinburne from St James's. We entered the laboratories and were met by Mr Hollingsworth who appeared to be very efficiently polishing and pipetting fluid into tubes. He then took charge of the visit and introduced the various technicians. It became obvious to us that this man was the technician in charge but then found to our amazement and amusement that he was in fact the chief bottle washer and phlebotomist. The laboratory was a small multi-disciplinary laboratory and produced microbiology, chemical pathology, and haematology tests, all by old fashioned manual techniques. The view of my colleagues was that the microbiology and chemical pathology should be moved down to the Infirmary.

There was no histopathology and further discussions would have to be held about the performance of post mortems, as this was important for recognition of training programmes for junior doctors. It was obvious that the most important discipline at Cookridge was haematology and a rapid turn round service was required as this was necessary for the treatment of patients with both radiotherapy and chemotherapy. It was quite obvious that a haematology service would have to be performed at Cookridge.

We had been despondent recently about the lack of space in the department at Leeds and the impossibility of me pursuing my major interest in leukaemia diagnosis.

However, here was a modern, well designed laboratory, spacious, airy, and with plenty of light, and with a sterile cabinet, and as nobody else wanted it, this laboratory would be for the sole use of haematology. I proposed therefore to purchase an automatic blood cell counter as part of the transfer of ownership as this would certainly release one or two technicians for other developmental work.

Tragically the Chief Technician was so terribly upset about the break up of his laboratory that he had a myocardial infarct and was very ill. He made a reasonable recovery but St James's Hospital agreed to take him on their staff.

I therefore had a vacancy for a Chief Technician, as well as two others. I was left with Hollingsworth, the washer up and phlebotomist, and another technician who dressed all in black, had long black hair, and big black sunglasses. He was immediately given

162

the nickname "the black shadow" by Infirmary staff and subsequently this became shortened to the nickname "Shad". He had a somewhat unusual personality and his extensive night life would be a problem to us for quite some time.

I therefore decided to advertise the vacancy for a chief technician to perform leukaemia diagnosis and research, as well as supervise the haematology laboratory. The first applicant had worked in the paediatric haematology laboratories at Sheffield. He took the job and did it very well but was missed in Sheffield and received an attractive offer to move back. I therefore had to advertise again and this time I got an application from a young man from Bath by the name of Steve Scott. At the interview it was appreciated that we both interviewed each other. I let him know what I wanted him to do and he requested that he be allowed to do what he wanted to do. Fortunately out aims coincided. He, in particular, wanted to work in leukaemia diagnosis and assess the value of Fc (the receptor for immunoglobulin antibodies on the cell) and complement receptors in the maturation of white cells. To my delight he took the job and throughout his time at Cookridge he completely exceeded my expectations of him.

I had therefore to get some scientific equipment for the laboratory. I requested an immunofluorescent microscope from The Leukaemia Research Fund and Professor Dacie, who evaluated the request, turned it down as I had not requested a sufficient number of high dry lenses which was absolutely typical of Dacie. I modified my request therefore and was successful getting this piece of equipment. I also asked for further items of equipment from The Friends of the Leukaemia Unit and obtained, among other items, a spectrophotometer and an Apple computer.

Other staff was recruited, in particular, a Senior Technician by the name of Howard Limbert whose life was mainly devoted to disappearing down large pot holes in various parts of the world, with his wife who was also an MLSO at York. Howard was responsible for finding a new series of caves in Vietnam.

Steve Scott turned out to be a very interesting young man. He had left school with O-levels and, as he was unhappy with his home circumstances, joined the army at a young age. In the army he found a post in hospital laboratories and at the end of his army stint joined the haematology laboratories in the hospital in Bath. When he arrived he immediately set up a whole range of tests, including a test for his Fc and complement receptors, and a full range of histo-chemistry tests in leukaemia.

In my work at the Infirmary I received many referrals from colleagues in the Yorkshire region but was never able to take them very far because of lack of facilities. Now I encouraged them to send liquid samples and slides to Cookridge Hospital where I wished to run a referral service. With the ever improving facilities referrals began to increase and many very interesting cases were seen.

For some little time children with acute leukaemia were given treatment in which both remission induction and maintenance were performed mainly as out-patients. However survival figures remained stubbornly the same. The Germans then introduced an aggressive new regimen known as the Berlin regimen. Many patients however needed admission during remission induction from septicaemia and some died. Overall, however, survival figures significantly improved, but some children with an initial good prognosis would have died in remission induction which was a major problem.

It was obvious that a new approach was needed in chemotherapy; diseases would have to be divided into sub-categories to see which groups would be benefited by a regimen and which groups would be harmed. In fact treatment might be tailored for the individual patient. I saw that many of these sub-categories would be determined by laboratory investigations and I envisaged that the department at Cookridge would play a major role in this. The Leukaemia Diagnosis Laboratory was therefore encouraged to develop and utilize new tests in the future. It would have to further develop histochemistry; use flow-cytometry with an ever increasing range of monoclonal antibodies; and enter the field of cytogenetics which progressively began to occur. This concept of right patient right treatment right time has progressed in medicine and is now known as stratified medicine.

Meanwhile, at the Infirmary, following Colin Bird's appointment as Professor, we arranged weekly meetings to discuss the interesting lymphoma cases referred to us. We had two successive senior lecturers in the Department perform the Lymphoma Diagnostic Service before they left and then Colin Bird advertised specifically for a Senior Lecturer with an interest in lymphomas and we were extremely lucky to appoint Ian Lauder from Newcastle who certainly knew more about lymphoma than anybody in Leeds. He made the weekly meetings very interesting and haematologists learned a lot about lymphoma diagnoses they had not heard of. He also was very important in the Lymphoma Therapeutic Trials confirming the true diagnosis on the cases entered into the various trials and placing them in the correct disease category.

Colin Bird and I thought it would be a good idea to apply to The Leukaemia Research Fund to find a major project that The Leukaemia Research Fund would finance. The Scientific Director of The Leukaemia Research Fund was a Mr Gordon Piller. Gordon Piller was previously an administrator at Great Ormond Street Hospital for Children in London. Parents of children with leukaemia had approached him as they wished to donate money to a charity for children's leukaemia. Such was the demand for this service that Gordon Piller founded The Leukaemia Research Fund and became its first Director.

When we approached Gordon Piller he told us that the policy in The Leukaemia Research Fund was to allocate various specialty research categories to different regions in the United Kingdom. Our major interest was in the diagnosis of leukaemia and lymphoma but he told us that Oxford was already designated as a Centre for the Development of Diagnostic Methods in Leukaemia and Lymphoma under the direction of Dr David Mason, a well known expert in the field.

In the Scientific Journal, Nature, in May 1964, a paper appeared written by W F Jarrett and co-workers in which they described a virus-like particle associated with leukaemia in cats. This became known as the feline leukaemia virus. It provoked enormous interest in the field of leukaemia and obviously the challenge was to find a similar virus in human leukaemia and lymphoma.

Piller's idea therefore was to designate Leeds and Yorkshire as a Centre for Epidemiology. The basic premise was that after epidemiologists had identified a leukaemia cluster, Piller would send a team of scientists from London to investigate this fully and look for the virus. Colin Bird and I would be the Grant Holders for this project and in due course an Epidemiologist, Dr Ray Cartwright, was appointed. Premises in Hyde Terrace were found, and Cartwright would hold a honorary position in the University Department of Public Health with Professor Richards.

Previous epidemiology studies on leukaemia and lymphoma had relied on hospital records or death certificate data. This trial was to be different in that the lymphoma diagnoses would be made by Professor Bird and Ian Lauder, and the leukaemia diagnoses by the laboratory at Cookridge, and myself.

For the Leukaemia Diagnosis laboratory at Cookridge this was marvellous news. All the cases in Yorkshire would be referred to Cookridge for accurate diagnosis and meant that an abundance of material would be available for research along other lines. A team of women was recruited. They would interview the patients collecting relevant epidemiological data, and also perform a phlebotomy on the patient and send a fresh sample to Cookridge, as well as bone marrow slides. Paraffin sections of the lymphoma would also be sent to the Department of Pathology for Professor Bird's attention. Thus haematologists and pathologists in the region received a modern up to date diagnosis, backed by the relevant histo-chemistry and immunological marker studies.

The epidemiological study was extensive and included some case control studies. These showed a weak association of lymphoma with atopic skin diseases and previous treatment with steroids and radiotherapy. Numbers however were small and not of major significance. The occurrence of Bell's Palsy as a precursor of lymphoma was confirmed.

Throughout the farming industry, records of fertiliser and pesticides used on crops have to be carefully recorded and all this data was available to the epidemiologists. They were, however, unable to find any correlation between agents used in the treatment of crops and the incidence of lymphoma or leukaemia.

Fears were expressed by the public in the country that electric pylons may be associated with an increase in cases of leukaemia. The Epidemiological Survey found no evidence to support these fears. Finally a map was prepared of the places in Yorkshire where leukaemia and lymphoma occurred but no clusters were found. In fact, the work continued in other regions in England until it was possible to produce a comprehensive map of lymphoma and leukaemia in England and this was published as a book by The Leukaemia Research Fund.

With access to all this biological material many papers were published from the laboratory at Cookridge and Steve Scott was able to prepare a PhD thesis with success and became a Fellow of the Royal College of Pathologists based on published work.

The work in the laboratory became more sophisticated, cultures of cells that produced monoclonal antibodies for detection of antibodies were grown in the laboratory and so significant sums of money were saved by the home - based production of monoclonal antibodies. The laboratory persuaded the Regional Health Authority to fund a FACS machine (fluorocein activated cell sorter) if the maintenance fees of this machine were paid for in the first place by The Friends of the Leukaemia Unit.

Meanwhile, as work on the Clarendon Wing was progressing the Heads of Department in Pathology looked to the future planning of

Pathology services in this building. The pathology laboratories were situated along one face of the new building. If Phase 2 had gone ahead, they would have communicated with further pathology accommodation in the new block. This was not to be, however, we were successful in persuading the planners to put windows in all the offices along this face which made them much more pleasant to occupy. However, the laboratories had no windows and were unpopular with technical staff.

It was obvious in the planning meetings that there was no link between the old and new buildings which would have made it very difficult for the transfer of staff, and patients, between various departments. It was agreed therefore to build a link, in the form of a tunnel on stilts, between the main corridor of the main Infirmary opposite the Medical School, which coiled round the Old Medical School to enter the Clarendon Wing. When it was built it was a highly effective means of communication but a visual disaster. It is said that the contract for the building of this link saved a steel contractor from going bankrupt, notwithstanding this he would not display any of his advertising boards because of the total lack of aesthetic appeal for this construction.

The Heads of Department at the Infirmary paid a visit to the Maternity Hospital to look at the laboratories there. The laboratories formerly situated at the Hospital for Women were now at Roundhay Hall with the exception of Blood Transfusion which was now settled within the Department of Haematology. I was amazed to find that red cell diameters were being measured in the laboratory of the Maternity Hospital. This was carried out by having two lights on a wall with a long bench underneath. The slide was then held up to these beams of light and moved along and above the bench until the two images of the lights coincided The slide is then lowered to the bench and the red cell diameter is read off. I, in my career, had vaguely heard of this technique but never seen it before, nor is it mentioned in the standard text book for Practical Haematology by Dacie & Lewis.

The technical staff however were somewhat dismayed by the thought of moving and joining another department. There appeared to be adequate space and they had a social room where they could play table tennis and other games and generally enjoy a relaxed life.

I remember meeting a young, pretty girl walking down the street by the name of Jane. I then made what I thought was a jocular remark and said "we will sort you out when we get you into our department". Her face dropped, she began to sob and, as I consoled her, tears poured from her lovely doe-like eyes. I do not know what any bystanders made of this but I suppose it was just an every day occurrence for the average consultant haematologist. Jane, however, did become the stalwart of the department and had a very fulfilling career as a technician and finally left with her husband, Richard Kendal also a technician in the department, to work in the United States. Richard had developed an assay for erythropoietin, prepared a PhD thesis, and was head hunted by an American instrument firm.

The Clarendon Wing finally opened in 1984 to much acclaim. Unfortunately on Day 2 the electricity supply from a newly built power station ceased and the entire building was plunged into darkness. The problem was compounded by the fact that many of the rooms had no natural light and the darkness would have been complete. All the lifts stopped, as might be expected, many were trapped including members of staff, patients, and the Dean of the Medical School. Fortunately this crisis has never occurred again.

All the pathology offices and laboratories were situated on D Floor. Haematology was allocated three laboratories and a donor area for blood transfusion which was now completely obsolete. I had plans that this donor area might become a one-day clinical area for venesection, transfusion, and administration of chemotherapy and I applied for the appropriate nursing staff. The senior nursing staff however remained completely recalcitrant to this idea.

There were two consultant offices and a small secretary's office close by. I occupied one consultant room and the adjacent was for an additional consultant haematologist that had been approved for the new hospital. The appointment of the third consultant in haematology was delayed so that local senior registrars could get their final MRCPath and apply for the post. The result of this was disastrous in that the Infirmary hit a financial crisis and the appointment of any new posts was cancelled. I was incensed by this news and went to see Richard Oswald, the District Administrator, who appreciated my point of view but in view of the financial circumstances could do nothing about it.

This financial position was resolved in time and Dr Derek Norfolk was appointed to this post of Consultant Haematologist. The Annual General Meeting of the British Society for Haematology was due to be held in 1983 and I was able to appoint Derek Norfolk's wife, Sue, funded by the British Society for Haematology, to act as the organiser for this meeting and the small secretaries office proved ideal for this purpose.

Along the D Floor corridor were situated branches of the other main departments in Pathology. There was a large exfoliative cytology section, a microbiology section lead by Peter Kite, and a chemical pathology section run by Alistair Stewart a non-medical scientist who was a geneticist. This was the first time in Leeds that all the main Patholog departments were working close together and this can have its advantages.

I got to know Alistair Stewart reasonably well and we began to discuss the new changes in DNA technology that were underway and were now being introduced into medicine. The specialty of haematology was one in which these techniques were being used, particularly by Professor David Weatherall at Liverpool University who had a major interest in the haemoglobinopathies. As I had very little knowledge of DNA I had to buy the relevant books and get down and re-educate myself. How I was going to introduce these new disciplines into haematology laboratory at Leeds I had, at this stage, no idea how to do it.

After the opening of Phase 1 it was not long before we were told that the Infirmary was taking over Chapel Allerton Hospital. Chapel Allerton Hospital was a small district general hospital situated mid way between St James's and the Infirmary and participated in acute

170

medicine, geriatrics, and rheumatology. There was a busy Out Patient Department. It was, at the time of takeover, under the charge of Peter Dosset, Chief Pathologist at St James's. Once again we were in the throes of re-organising a service suitable for the hospital and according to our practice and principles. I insisted that there should be a haematology laboratory with an automated counter at Chapel Allerton and that it would probably be necessary to run a Haematology Clinic there to look at some general haematological problems and maintain an anticoagulant service. This again produced its management problems in dealing with the existing staff who had affiliations to St James's, and re-organising them to our pattern of work and also assimilating into the main laboratory from time to time to keep up standards. It also meant that our existing staff in the laboratory would have to rotate to Chapel Allerton as the service demanded.

Thus there seemed to be one management problem after the other and it seemed timely in 1987 that the Regional Medical Officer in Harrogate sent for me in my role as Regional Advisor in Pathology and told me that the Department of Health was asking Regional Health Authorities to provide management training for doctors. The Region had already appointed Helen Jones, a Management Development Adviser, who had been a teacher, counsellor, and partner in a Psychology Training Centre, and was an independent Management Development Consultant. She had participated in a recent massive re-organisation of British Airways which was universally praised. The theme of her work was a personal construct psychology.

I had discussions with Helen and also Martin Rogers, who was a Director of College Short Courses for the Region. He had been closely associated with GP Training Schemes in the Oxford and Yorkshire regions. After discussion we agreed that there would be a series of short residential courses lasting some two to three days at Highfield House, which is part of the College of Ripon and York St John, in Ripon. I proposed that each group should consist of a consultant and a senior registrar from each of the four major disciplines and, of course, the opportunity to join the training courses was open to all consultants and trainee senior registrars in the Region. The first course was held in November 1987. We soon learnt to work in small groups to solve problems put to us and this would lead to these small groups in separate rooms coming to conclusions and putting them down as headings on a series of flip charts. The groups would then meet together and present their conclusions from the flip chart, which everyone could see, and these would then be assimilated under the supervision of Helen Jones. The main theme for manager training we decided upon was that of The Management of Change and which would involve many of the problems I had encountered with the merger of the Infirmary with other small hospitals, and a close relationship with Wharfedale Hospital. The hypothetical change we decided upon was a purely imaginary merger of Wharfedale with the Infirmary. In real life this would have been political dynamite. After several two to three day sessions we finally came to some conclusions and recommendations, and these were published as a small booklet.

The meetings were enjoyed by all, in particular the senior registrars who found throughout these activities everyone had the same status and consultants did not tell senior registrars what to do. A further aspect to the course was the personal development of individuals which mainly took place in the form of groups tackling a problem and then meeting altogether to provide the solutions.

One particular exercise would affect me personally and lead to me taking a complete reassessment of my life and my view of myself. At the time there was a system of management devised by a man called Belbin. His management theory was based on management by groups or teams but the important thing about these teams was that they have to be carefully built with members of very different personalities.

The core members of such a team were given names which are not always easy to appreciate unless they are explained. Some examples are as follows:-

172

Chairman:
He has to be a stable, dominant, extrovert personality. He has to have charisma, be a good talker and excellent communicator.

Company Worker:
A stable and controlled individual being practical and disciplined whose job it is to work out the strategies into manageable tasks for each team member. A very systematic individual.

Shaper:
This person has an anxious, dominant, extrovert personality. Likes to lead and is the person who actually gets things done and implements group policy.

Plant:
This individual is dominant with a very high IQ, introverted, capable of being thrustful and uninhibited. Can be prickly. He or she is the ideas person. He or she provides the team with original ideas, suggestions, and proposals, and takes a radical minded approach to problems.

Monitor/Evaluator:
This person again has a high IQ. Is a stable, introvert, serious "cold fish", critical, rarely enthusiastic. This individual provides measured and dispassionate analyses of the various problems that are about and, in actual fact, is probably the corrective for the wild ideas that the Plant may produce from time to time. He balances the Plant.

Course Investigator:
Is a stable, dominant, extrovert, sociable and gregarious. Needs stimulus by others but, in fact, usually turns out to be the salesman for the group.

Team Worker:
A stable, extrovert, unassertive, likeable person who is there to produce unity and harmony among the group. A builder rather than a demolisher.

Finisher:
This is an anxious, introverted individual, obsessed with detail. He has great self control and strength of character. He is a compulsive "meeter" of deadlines and a fulfiller of schedules. He has a permanent sense of urgency and gets things done.

I had read all about the Belbin System of Management as it had appeared in the newspapers and magazines. I had decided without doubt that I was the "Chairman" as, in fact, as Head of Department that appeared to be my major role. Looking at all the other personalities and reading about them I appeared to take an instant dislike to the "Plant". There had obviously been illustrations in some of the magazine articles which did not help so I did conceive that in any organisation I would have difficulty in dealing with the "Plant.

It was much to my amazement that when I filled in a questionnaire with the others at a session in Ripon that I found out that I was a "Plant" with some of the characteristics of a "Shaper". I must admit that I had some difficulty with this at the time but it did lead me to read round this to find out more about the personality of the "Plant". I soon began to appreciate that a "Plant" is often a person with strong right brain characteristics. The left side of the brain deals with logic, mathematics, music and speech. The right side of the brain has more to do with assessment of space and visual concepts. It is said for example that these characteristics are the main components of a successful architect.

As far as I was concerned my main strength as a haematologist was in making diagnoses using a microscope and I had a very good visual memory. This reinforced the idea that I had a strong right brain. As I began to read round more of the subject I began to appreciate that this accounted for some other problems I had in life. I have a mild dyslexia and to look at any of my essays or letters you will find numerous crossings out as often the letters come out in the reverse order. I found to my complete surprise that when I tried to write with my left hand I wrote mirror image. Another problem I have is that I am unable to tell my right hand from my left hand. When I took my driving lessons and test this often descended into farce as I often went the opposite way to which the instructor told me. It also affected my participation in sport as I have no dominant side and as I ran to vault when I did physical education I did not know which leg to jump off. I also played a lot of cricket, and later golf, sports which are sideways sports. At cricket and golf I play as a left hander, which appears natural to me, but it means that the bottom hand, as I grasp a cricket bat or golf club, is the weak one, so I could never hit a cricket ball hard, and when I hit a golf ball I invariably sliced it.

Thus my whole being and personality was revealed unto me and at my stage in life this was a profound shock although it explained very much about me and my physical and mental characteristics.

I then began to understand many of the problems my colleagues would have with me because there was no doubt that as a "Plant" I was capable of developing new ideas and concepts and, with my "Shaper" characteristics and being the Head of Department, there was no one to stop me with my wild ideas. I also made decisions which were intuitive and not logical and I obviously shocked Charlie Buchan, the Senior Chief Technician, from time to time as I would walk into his room and completely out of the blue come out with a new policy decision which appeared to be totally illogical at the time. In my further role in the Institute of Pathology, which is the subject of the next chapter, I would again come out with proposals which colleagues would be very upset about because I had not undertaken SWOT (Strength, Weaknesses, Opportunities, Threat) analysis. Again, to sum me up, I am a person who took the broad outline of a problem and did not worry too much about the detail.

This training was very valuable to me because I did not know it at the time but I would finish off my career in medicine with a major management role within the Leeds Western District.

At this time there was a movement to good management throughout the country and many management firms sprung up producing all sorts of systems. For example, it was not uncommon to find a firm that would allow you to enter baths of fluid at room temperature with a specific gravity which allowed you to float and lose almost a sensual contact with the outside world and thus provide complete relaxation, and there were numerous books published some of which had bizarre titles, for example, one book on management was called "How to Teach Elephants to Dance".

I was impressed by two authors, Peter Drucker who had made his name as a Management Consultant in Japan after the war and can be given some credit for the post war Japanese Industrial Revolution.

Tom Peters was a messianic teacher of good management and ran day courses up and down the country and in the United States as well. From Peter Drucker I developed two main themes. One was that increasing specialisation was inevitable in a rapidly changing world, and secondly that these specialists worked best in teams. This I accepted as my main theme for the way I intended to manage the Department in future. I thought at first that this was quite original but then as I thought about it more I realised I was reverting to my youth when I played a considerable amount of cricket and my vision of management was that of a captain with a cricket team. I did however put this idea into practice and with success.

The second thing I got was from one of Tom Peters's books. It was one, that staff should have ownership of their job, and two, the potential for development of any individual is limitless. The extension of this is that I have an intense dislike for line management. My view is that most workers in the Department know exactly what they are doing, they do it well and do not need anybody to tell them what to do. As far as I was concerned, for the most part, they were experts who knew far more about their own individual job than I did. So the concept of giving workers ownership of their own job is that they should not have people directing them what to do, when they know already. Instead individuals should be encouraged and given support. As far as the development of individuals was concerned I considered that we were well enough staffed to allow individuals to pursue projects of their particular interest and, if appropriate, work toward a higher Degree. This policy succeeded beyond my wildest dreams.

Following the upgrade of the side ward on Ward 29 in 1982 the clinical practice continued to grow. But conditions were still unsatisfactory for patients with acute leukaemia were still being treated on the general ward. For the management of out patients requiring day treatment with chemotherapy or blood transfusion, in - patients had to vacate their beds to allow day patients to lie down for treatment. The quest therefore for better accommodation for Haematology continued and in 1983 the Clarendon Wing opened and bids were made for wards there, but the paediatricians had priority and all bids from Haematology failed. However, on D Floor next to the pathology laboratories, there was a room designed for blood transfusion donors. The concept that blood transfusion donors would turn up to a hospital ward to be bled for transfusion was now a completely out of date concept and this room was now obsolete. It did however present a completely suitable out patient treatment suite for Haematology. There were couches, toilets, a kitchen nearby for making tea and other refreshments, so therefore we applied for nursing staff to run it.

However the nursing hierarchy would not move from their entrenched position, they considered that it was not possible to treat a patient during the course of one day and that all patients who entered the suite would be admitted for overnight accommodation. The bid therefore failed yet once more.

Eventually a place was offered on D Floor, Brotherton Wing, formerly private patient accommodation, with single rooms and four-bedded rooms. This was entirely suitable accommodation and was a very big boost for morale in Clinical Haematology. Fourteen single bedded cubicles, including two sterile cubicles, were provided. This would, of course, need a great deal of money but The Friends of the Leukaemia Unit took up this challenge, had a major appeal, and raised £330,000 for the upgrading and provision of equipment for this ward. The large balcony attached to the ward was also enclosed in with glass and created more space which helped with the day patients. It was with triumph in 1993 that Lord Harewood, who had agreed to become Patron of The Friends of the Leukaemia Unit, came to the Infirmary and opened the ward. As more reorganisationsin the Infirmary took place, the Orthopaedic Out Patient Department located in the National Health Service side of the Brotherton Wing, became available. It was immediately opposite the Haematology In -Patient accommodation and thus created a united Department. Inside the new unit would be rooms for harvesting stem cells, one day facilities for treating out patients with chemotherapy, and blood transfusions. There would also be offices for consultants and clinical secretaries. Being situated in the Brotherton Wing, it was almost above the main haematology laboratories in the basement of the Martin Wing.

Again, The Friends of the Leukaemia Unit, now called the Friends of the Leukaemia and Lymphoma Unit, provided much of the finance. £157,000 was raised by The Friends and it was opened by the Calendar Girls in 1993. The Calendar Girls then contributed £50,000 to The Friends of the Leukaemia Unit.

A problem we had in Clinical Haematology was that we never seemed to have enough junior staff from the Regional Training Scheme.

However, early in our history, Dr James Lynch, the Post Graduate Dean, asked me if I could take young Sudanese doctors for haematological training. I was very willing and the first trainee was a young woman by the name of Anwar Kordofani. She was the daughter of a Lecturer in the University of Khartoum affiliated to a University in London, and spoke fluent English. In fact, as a little girl, she had presented the Queen with a bouquet of flowers when she visited the Sudan on a Royal visit. She was bright, intelligent, an apt pupil, and got through her MRCPath with no difficulty. She was the first of several Sudanese doctors, who all spoke excellent English, were a great help to us, and were successful in their examinations.

Eventually Anwar wrote to me to say that they now had sufficient trained haematologists to run their own training programme and would I go to the Sudan to act as an external examiner. This I agreed to but the Sudanese President imposed such a strict Muslim regime that this proved impossible.

We received several trainees from third world countries for visits of one year or more, some of them backed by the British Council, and many of them were very interesting individuals. I soon learned of the poverty in these nations and that what these haematologists would learn would not be put into practice back home because there was no money.

Three of our visitors, once they had got to know their way about, disappeared. I subsequently found out they had got registrar jobs; which for them, provided almost unbelievable wealth. Some years later I received a phone call soliciting advice on a clinical problem. I recognised the voice of one of our absconding registrars who was now a locum consultant! He would be receiving a salary that he could never have dreamed about.

We also received visitors for two or three months. I remember one middle aged haematologist from a central African state. I proposed to take him to the Regional Blood Club where interesting cases are presented and discussed. This meeting was at York and so I thought I would take him for a ride and show him some important historical sites. We stopped at Towton where the deciding battle of the War of the Roses was fought and 30,000 lives were lost. To my total amazement he had extensive knowledge of the extended family of Edward III and the major participants. He must have been taught this as a child by an English teacher. His knowledge of English history was far in excess of any school child of today.

Our visitors contributed much to the Department and they taught us a lot about life in less fortunate countries.

Modern chemotherapy of leukaemia and lymphoma began in the Royal Marsden Hospital and St Bartholomew's Hospital in London. The medical staff who were pioneers in this form of treatment were physicians who, in due course, were called Clinical Oncologists. Though the Clinical Oncologists were pioneers in the treatment of leukaemia and lymphoma, the treatment of these disorders was mainly taken up by haematologists in the rest of London and the United Kingdom.

Oncologists began to treat solid tumours and so a new specialty of Clinical Oncology had therefore been created. There was a problem however in that the clinical oncologists and the haematologists basically were interested in treating the same tumours because these were the ones that responded best to treatment.

The Cancer Research Campaign continued to pursue the appointment of clinical oncologists throughout the country with their financial support and in 1987 The Cancer Research Campaign proposed to the University of Leeds, and the Leeds General Infirmary, that they should appoint a Professor of Clinical Oncology to be situated at the Infirmary. A named individual was proposed and unfortunately his publications indicated that he had the same interests as Dr J A Child , much to his dismay. Dr Child was, at that time, Chairman of Faculty, the main Medical Advisory body of the Infirmary, and through this body he mounted a campaign of opposition. The District Administrator was worried about the cost of chemotherapy that a new Clinical Oncologist would engender. There was therefore a stalemate in the situation.

The District Administrator at St James's, however, had imagination and flare and was known to be an entrepreneur. Despite the similar financial disadvantages at St James's, he offered to take this new Chair in Oncology and the Cancer Research Campaign, and the University of Leeds, soon agreed. In 1988 Dr Peter Selby was appointed to the Chair of Cancer Medicine at the University of Leeds, and was given beds and out patient facilities at St James's Hospital. Professor Selby did not antagonise the haematologists and went on to build an excellent Department and became a national leader in the field. He became the right man in the right place when the Bexley Wing, a regional centre for oncology, was built.

Going into the 1980's, Heads of Department in Pathology, Professor Lathe, Chemical Pathology, Professor Mary Cooke, Medical Microbiology, and Professor Colin Bird, Histopathology, and myself, faced a world of ever increasing complexity. Our departments were about to amalgamate several other small hospitals in Leeds and many staffing changes. The Royal College of Pathologists was newly created and was organising exams which would involve all of us, there would also be various college committees, such as Standing Advisory Committees in each speciality. All of this would be time consuming.

There were, during the next ten years, several changes in the professorial establishment in Pathology. Grant Lathe, a distinguished scientist, retired and was replaced by Dr Brian Morgan. During the 1960's Professor Lathe had a bright idea that the subject of Chemical Pathology should be more clinically orientated and he appointed a distinguished clinician, Paul Fourman, to this role. He appointed Brian Morgan as his Senior Lecturer. They both then enthusiastically investigated many cases of patients who had had a partial gastrectomy and examined them for evidence of a bone disease.

To do this they performed a trephine biopsy, that is, they took out a core of bone from each patient and examined the bone structure. They accumulated a good deal of material but, unfortunately, Paul Fourman collapsed and died of a myocardial infarction in the corridor of the Medical School.

Brian Morgan returned in the 1980's as Professor of Chemical Pathology and continued his interests in metabolic bone disease.

Mary Cooke moved out to London to work for the Public Health Laboratory Service (PHLS) when her husband, who was a senior figure in the administration of the law, moved to take a more senior post in London. She was replaced by Professor Richard Lacey, very much a controversial figure. He had left Bristol University in controversial circumstances and went to work at Boston in Lincolnshire.

After the appointments committee I remember that it was decided that Professor Lacey would be a lively figure and someone like him was necessary to liven up the Medical School. Professor Lacey not only livened up the Medical School, but also the Infirmary and numerous supermarkets in Leeds.

His main interest was in the microbiology of food and he went around with his lecturers visiting supermarkets, opening their fridges, and taking the temperature. If there was anything wrong with the fridge then the public were soon made aware. It was not long before Richard was banned from entering quite a few supermarkets around Leeds.

If one wishes to know more about Richard Lacey, and uses Google, the information about him starts with the heading HOW NOW MAD COW. There was in England an outbreak of mad cow disease, known as bovine spongiform encephalopathy. He found that the cows developed this disorder from eating prepared food which included chopped up meat from cows which was obviously infected. This caused an enormous crisis in the agricultural industry and all UK beef products were banned in the EU countries. The use of these animal food products which had obviously transmitted the disease was banned, but Richard Lacey entered the fray by suggesting that the agent causing this disease could be transmitted to humans through eating infected beef. There is a human equivalent of mad cow disease, known as Creutzfeldt- Jakob disease. This is a fatal disease but has been rare in Great Britain. As a result of Richard Lacey's propaganda, which filled all the newspapers, the Government which had been slow to act ,eventually legislated that no cow carcass must be sold until the brain and spinal cord had been removed. Richard Lacey however still remained controversial. He was a Don Quixote figure who loved to tilt at the establishment instead of windmills.

There was in the middle of this decade some movement in terms of accommodation. Some scientific microbiology moved from the Old Medical School up to the Science Block in the University freeing up some space. Brian Morgan considered that it would be advantageous to centralise the Department of Chemical Pathology and therefore vacated the Martin Wing.

There was also the dreadful shortage of space in Haematology. The Department then moved out of its accommodation opposite the Nurse's Home and down into the basement of the Martin Wing. Far more space was available but it needed a good deal of work to be done to upgrade it and to convert it for use as a haematology laboratory. Air conditioning was introduced and finally Haematology had some high quality purpose built accommodation and, despite the fact that the Department was in a basement with just a few windows looking up to the road above, the staff were well pleased with the changes. Blood Transfusion then returned from temporary accommodation on the Princess Mary Ward to fill the vacated haematology laboratory.

Unfortunately, not too long after this, Brian Morgan developed a highly malignant tumour, the origins of which were never found, and it consumed him in a very short time. Brian Morgan was replaced by John Whicher. John Whicher had an international reputation in the field of the immunological factor complement and was very good at simplifying the subject and giving very good lectures. In previous years the Department of Chemical Pathology had taken over the

Department of Immunology when the existing Professor, Professor Gerry Gowland, moved on. Finally in 1986 Colin Bird moved back to Scotland. It is often said that the Scots send their bright young men and women down to be professors in England and to make their mistakes on the English until they are the completed article, in which case they then return to become Scottish professors. This is what happened to Colin who moved to take the Chair vacated by Sir Alistair Currie, who was Professor of Pathology in Edinburgh, and master of an enormous pathology empire. This post was tailor made for Colin and he went back to Scotland to run it with the same ruthless efficiency as had his previous master. Colin subsequently became Dean of Edinburgh Medical School.

Before Colin Bird left for Scotland we were devastated to learn that Ian Lauder who provided an excellent diagnostic service in the field of lymphoma had been invited to take the post of Professor of Pathology in Leicester. Ian's reputation had risen with his work in Leeds and his contributions to the Epidemiological Survey. The post was advertised and was taken by Dr Andrew Jack from Glasgow.

We were very fortunate with this appointment. Andrew, as well as being an excellent diagnostician, had a sound scientific background. He had undertaken an intercalated year at Medical School and been awarded a PhD based on a research project. When appointed, he organised the processing of trephine biopsies (cores of bone marrow obtained from the patient and processed histologically as opposed to bone marrow aspirates which are spread on a slide and stained like a blood film) so that they were mounted in plastic instead of fixation in formalin and mounted in paraffin wax. This presented a whole new field of study. The cell appeared to have a different morphology, one could see the architecture of the bone marrow and use monoclonal antibodies to identify the cells. This service was available in very few other Centres.

Andrew became an intellectual leader for the haematology team and this story will be continued in the next chapter.

Professor Bird was replaced by an American working at St Mary's Hospital Medical School in London, by the name of Art

Boylston, who hailed originally from Boston, Massachusetts, in the United States. Art Boylston was a pleasant extrovert character and was appointed on the basis of his having a research grant with the Medical Research Council which counts very highly in the research assessment exercise.

The Medical School was now reaching a point of financial crisis. The financing of the Medical School from the Government is made on the basis of a regular research assessment. Finance is based on research produced, publications in internationally recognised journals, and research income. Leeds was not doing well, it was becoming impossible to become a professor and head of a service department and prosecute serious scientific research. Art joined the Department of Pathology and appeared determined to reorganise the routine work to release funds for research but countered a good deal of resistance.

The Dean of the Medical School took a stringent look at all Medical School finance and the possibility of future development seemed poor without outside finance. Monty Losowsky, Professor of Medicine at St James's, took action on this. There was a research fund known as The West Riding Medical Research Trust which had been largely founded by Dr Chandler at Chapel Allerton Hospital. He used this money to create a Chair of Molecular Medicine who would direct a Molecular Medicine Unit at St James's. Professor Alex Markham was appointed to this Chair and he developed this Institute into an internationally recognised Centre for molecular research.

However there were very drastic changes undertaken in the School of Clinical Medicine in the Medical School. From 1 October 1991 all departmental structures were abolished. This was to be replaced by three research institutes: Laboratory Sciences; Physical Sciences, and the Institute of Epidemiology and Health Service Research. It was obvious that the University Pathology Department would belong to Laboratory Sciences and all relevant personnel and equipment would belong to this Institute. Thus the Heads of

Pathology and Chemical Pathology would move to this Institute and take personnel and equipment with them. Thus Professor Art Boylston and Professor Whicher left for St James's and began to asset strip their departments. The Department of Medical Microbiology was part of a much larger department which was a member of the Faculty of Sciences in the University.

What was absolutely certain is the likes of Professor McNicol and Professor Bird, omnipotent Heads of large departments producing service work as well as research would not be ever seen again. Thus in the late 80's and early 90's there was a state of academic turmoil in Leeds. One of the consequences of this was that the University of Leeds laid down an ultimatum to the Infirmary to say that the amount of money they were paying for the pathology services was very inadequate and the Infirmary must pay more money. The problem was the Infirmary had no money and the services they were paying for had radically changed. The consequences for the pathology service is the major topic of the next chapter.

CHAPTER 11

THE INSTITUTE OF PATHOLOGY

"Render unto Caesar the things that are Caesar's and unto God the
things that are God's"
Mark 12:17

In 1987 Margaret Thatcher was re-elected for the third consecutive
time. She approached the reforming of public services with maniacal
zeal and health was no exception. The new Secretary of State for
Health, Kenneth Clark, set about introducing the concept of the
market to the National Health Service. As they said at the time "the
money will follow the patient".

This meant, in essence, that the National Health Service would be
divided between purchasers and providers, and in Pathology there
was no doubt we were providers of a service. Thus various
administrative bodies were set up and general practitioners, certainly
in group practice, were invited to become Fund Holders. They
would receive a budget and pay the hospital for services every time
they referred a patient. This meant re-organisation of the Practices
and for this purpose Practice Managers were appointed. This was by
no means compulsory for general practitioners and some single
handed general practitioners remained.

There was an administrative body in the centre of Leeds, which
was known as the Leeds Health Authority, which also purchased
services from the hospitals in Leeds. Hospitals would in future be
run as Trusts. A Trust was set up at the Infirmary to run the Leeds
Western District with Tony Clegg, a multi-millionaire property
developer, as its Chairman, and Stuart Ingham its Chief Executive.
Len Wright was the Finance Director. Tony Clegg turned out to be a

pleasant individual who was on the whole liked by the majority of staff.

There were worries with the new system. I, in Pathology, was a little worried that the GP's might decide to receive their pathology services from hospitals other than the Infirmary, but this never happened.

Just after the formation of the Trust, the University of Leeds requested more money from the Infirmary for pathology services. There were no problems with Haematology because this was fully NHS in its financing but clearly there were problems with the rest of Pathology. All of the previous agreements on the so-called knock-for-knock basis had been agreed without very little recourse to accurate statistics and there was no way of knowing who was right in their assessment of the funding of pathology services.

The Infirmary therefore decided to fund its own pathology services, including histopathology, chemical pathology, haematology; the University Department of Microbiology through its Head, Richard Lacey, refused to have anything to do whatsoever with this proposed new body.

I found myself in the post of Director of Pathology but no one ever discussed the terms and conditions of this or even asked me to do it. It was just generally assumed that I was, as the Senior Pathologist, the one to direct the new organisation. Professor of Chemical Pathology, John Whitcher, and Head of Histopathology, Professor Art Boylston, had departed the Infirmary and joined the Laboratory Sciences Research Institute at St James's. The Head of Histopathology was therefore represented by Dr Philip Quirke, Senior Lecturer in Pathology, Dr Ian Barnes, a top grade non medical scientist, and Haematology by Dr Derek Norfolk as I had moved up to be Director of the Institute. The Department of Haematology had little to do, but the Head of Histopathology and Chemical Pathology had an enormous task.

They had to work out how much of University space was used for hospital services, and all senior lecturers who performed routine hospital services, and all technicians performing hospital services, were moved over to the employment of the NHS. Thus the Hospital paid for rental of University space and employed all the staff performing hospital services. The question was whether the Hospital was paying too much to the University for its services and I think the answer at the end of the day was, no. Through all these deliberations we had the help of our Institute Finance Officer, Philip MacDonald, who was extremely good and helpful. He had an assistant to help him but that is all. We also were awarded a Personnel Officer who used my secretary. Thus we had a total management staff of three, in fact the individual departments were managed by the medical or scientific heads of department, and the routine services by senior chief MLSOs all of whom were extremely good at managing budgets. Philip MacDonald worked closely with all the senior chiefs in managing the finances of the various departments and made sure that they did not overspend.

The management body which ran the Institute of Pathology was a Management Board the Chairman of which was Stuart Ingham, the Chief Executive of the Trust, and Len Wright, the Finance Director of the Trust. Also on the management body was the Director, plus Heads of Department. We also had regular contact with a middle manager of the NHS who gave help in many of our other management duties.

The structure of the Management Board was modelled on that of a Trust Board and two non-executive Directors were appointed. One was the Company Secretary of ASDA, and another one the Personnel (Human Resources) Director of Northern Foods. The majority of problems brought to the Trust Board for decisions were those made at its subsidiary body, the Heads of Department. It was again a body comprising the Heads of Department, a middle grade NHS Manager, Philip MacDonald, and the Personnel Manager.

In the Institute we had considerable freedom to do what we liked. For example, if we needed a new consultant then the Head of Department would have to find the funding by economies within the Department, get it approved by the Heads of Departments Meeting and then present the Institute Board with a fully documented case of need.

There was another monthly meeting which concerned me every month and this was a meeting with the Chief Executive, Stuart Ingham, and Finance Director, Len Wright, to discuss the monthly budget statement. This they greatly relished, looking forward to tearing me to shreds and I was also apprehensive as I never counted budget management as one of my special skills. However Philip MacDonald produced the statement and I went through it with him in great detail until I had a full understanding. What helped matters very much was the fact that after the first month when the senior chief technicians bought reagents and instruments in bulk to achieve maximum discount then the Institute never overspent. We had a cost improvement charge of somewhere between 2% and 3% to find and this we did, and after five years of being Director of Pathology the Institute of Pathology never overspent and, moreover, a similar parallel Institute of Dentistry run by the dentists also never overspent. It appeared the best financial managers within the Trust as a whole were those of the Institutes run by professional people.

At the beginning of my job as Director of Pathology there was one very dark moment. I was asked to interview, with the Personnel Manager, a non-medical scientist by the name of Chris Chapman. Recently Dr Clive Hayter, the Head of Nuclear Medicine, had died and the decision was made to close the Department of Nuclear Medicine and divide its various services among other departments. For example, the imaging service went to Radiology, and isotope studies in haematology went to the Department of Haematology where a technician by the name of Richard Kendal, a very able young man who had trained in the Department of Nuclear Medicine, was now able to take over the isotope studies from there with no problem. Whilst in Nuclear Medicine, Richard Kendal was able to pursue a PhD on Erythropoietin Measurement in Health and Disease. To do this Chris Chapman helped him to raise an antibody against erythropoietin in animals and label it with a radioisotope. This could then be used in a radioisotope dilution technique to estimate the erythropoietin level. There is no doubt therefore that Chris Chapman had very special skills within the Department. He belonged to the Department of Nuclear Medicine, which was known within the Hospital for its strong left wing views, and it was so red as to be "incandescent"; Chris Chapman was no exception.

At the dissolution of the Department of Nuclear Medicine he was moved to the University Department of Chemical Pathology which was a disaster. The culture change for him was enormous. He moved from a department with a great degree of freedom in managing its own work to an authoritarian professorial regime.

He became very unhappy and began to make a series of accusations against members of the University Department. Colin Toothill was a non-medical scientist who worked in the Hospital Service Department. He was very popular in the Hospital and helped many individuals with their research projects and getting techniques to work. In particular he spent time in the Haematology Department helping Philip Day in the field of red cell biochemistry. I saw a lot of Colin Toothill and regarded him as a personal friend. When the Infirmary took over Chapel Allerton Hospital space was identified for Colin to set up a Lead Assay Service for other hospitals and various industrial concerns.

This produced considerable income and he was accused of paying some of this into a university numbered account from which he ran Christmas parties of great renown. I had a University numbered account which was very useful, in particular, when we were running the Epidemiological Survey; then the Leukaemia Research Fund paid money into my numbered account which funded the diagnostic work for this project. Unfortunately for Colin Toothill this was deemed not to be sound practice and he lost his hospital contract. This was a major disaster not only for Colin but for all his many colleagues.

Chris Chapman joined the research group of Professor Whitcher and participated in the various research projects taking place ; a paper resulting from one of these projects was submitted for publication to the magazine Nature, one of the leading scientific journals in the world. This was accepted for publication and at this stage Chris Chapman wrote to the editor of the journal to say that Professor Whitcher had manipulated some of the laboratory results to suit his overall conclusions. This was a potential career disaster for Professor Whitcher.

Ian Barnes, a top grade biochemist, who was responsible for the Hospital Service, had declared his intention of not using the home made kits for immuno-assay of thyroid function which Chris Chapman had produced, but buying kits from a commercial firm on the grounds of economy. Chris Chapman was very upset by this and made the accusation that Ian Barnes was taking backhanders from the producer. This again was a grave insult which could have led to the sacking of Ian Barnes.

He was reported in the first place to the Personnel Department who warned him that if he should make further false accusations against anybody he would be sacked. He did, but he was not sacked. Instead the Department of Chemical Pathology with legal advice decided to institute a re-organisation programme of its work. It was no surprise that at the end of this exercise Chris Chapman would be found to be redundant.

On the first day of my appointment as Director of the Institute of Pathology I was asked to accompany the Personnel Manager to interview Chris Chapman. He was therefore told of the re-organisation plans, told of the fact that there was no place for him, and that he would be made redundant. This decision was made on the day before his birthday which meant that Chris Chapman would lose his pension because he had not reached the correct age. Chris Chapman's union representative was present and both Chapman and his representative showed great resentment, as might be expected.

The end result of this was that the union representative made various statements to the press and before long Chris Chapman was known as the "whistle blower" and his story was published in all the main major daily newspapers and was reported on the major news bulletins. The Labour MP's of Leeds took up his case and it was discussed in Parliament. The furore went on for quite some time and finally it was decided to ask Lord Merlyn Rees, a former MP in Leeds and, in the last Labour government, the Home Secretary, to look at the documents and make a report on the accusations against Ian Barnes. After a few months he did make a report and this exonerated Ian Barnes completely.

This was for the Infirmary as a whole a major disaster. This was a disaster that I foresaw when the Department of Nuclear Medicine was set up. Radioisotope tests are part of the normal complement of tests that a service laboratory such as Haematology and Chemical Pathology might produce and in particular members of staff in these departments might need to have training and expertise in these techniques for their professional exams. It was inevitable therefore that conflicts would arise but these were largely avoided by the fact that Clive Hayter was not only a very powerful individual but a very personable one as well. Chickens however came to roost when the Department was split up and I feel at the end of all this that Chris Chapman, though he should not have made these career threatening accusations against his colleagues in Chemical Pathology, was badly managed and he should probably have been sent to another NHS department.

The departments of the Institute relished their freedom and some began to attempt to earn money outside the NHS. A group of doctors and technicians went over to Saudi Arabia to see if they could produce investigations for the Saudi hospitals over there. They ran foul however of the fact that they had to pay a representative who would then apply the various bribes to the various organisations concerned. The difficulty of earning outside money was seen in the Haematological Malignancy Diagnosis Service (HMDS) *(see later)* which made its own monoclonal antibodies for diagnostic purposes. To be able to sell these they had to purchase a license for manufacture and a separate license for selling this product. The cost of the licenses was prohibitive.

When the Clarendon Wing opened I moved my office from the Department into D Floor, Clarendon Wing, in view of a dire shortage of space for the secretaries. One virtue of working on the Clarendon Wing was that one met and got to know members of the various other pathology departments situated there. One such person was Alistair Stewart, a non-medical lecturer in Genetics, from the Department of Chemical Pathology. He was very excited about his current work in recombinant DNA technology and repeatedly told me that the Department of Haematology should be into this subject. There were problems however in that there appeared to be no suitable member of staff to train and the equipment was expensive and would be difficult to obtain in the usual way.

By chance one day I dropped in to the technician's tea room in the basement of the Martin Wing and heard a young technician by the name of Paul Evans giving a dissertation to his colleagues on the structure of DNA and techniques used for its investigation. I sat down and listened to him and was amazed to find that he appeared to have a very good grasp of the situation. He had obviously received some instruction from Alistair Stewart but had also read widely into the subject. Here was the opportunity I was looking for, I took Paul over to see Alistair Stewart and after discussion Paul was sent to the University Department of Genetics for training for several weeks. When Paul returned he presented me with a list of equipment which received almost immediate funding from The Friends of the Leukaemia Unit. We had an under used laboratory in the Clarendon Wing which was entirely suitable and so within a few short weeks we were up and running.

I had recently negotiated two more senior registrars in Haematology for the Regional Training Scheme, despite the fact that virtually all my colleagues in the region wanted them as work horses I was determined that each Senior Registrar should have the opportunity of taking a year out to pursue some research project as part of their training, usually after obtaining the final MRCPath.
One such Senior Registrar was Caroline Shiach. I already knew that Caroline, who had recently obtained her final MRCPath examination, had undertaken research for a Doctorate in Medicine in Scotland before she came down south. She had all the material for an MD but had never written it up. I therefore seconded her for a year to the newly created DNA Department to help Paul and write up

her MD thesis. I told her that I would stand over her until she accomplished this task which she duly did.

There was then a major exercise in regrading MLSO's in all disciplines throughout the country. Each technician was reviewed separately and told their new grade. Paul was an MLSO Grade 1 which is the basic grade, but was quite surprised when I promoted him to MLSO 3 which was the grade of a Head of a Section.

I also took the opportunity of involving Andrew Jack who had been appointed in 1986 to replace Ian Lauder who had obtained a Chair in Leicester. Andrew had an intercalated degree and a PhD from Glasgow and was well versed in science and the techniques involved in recombinant DNA technology.

He invited Professor Goudie from Glasgow to talk about his recent work on the immunoglobulin gene. Recently Professor Alec Jeffreys in Leicester described a technique of finger printing DNA which also involved the use of the polymerase chain reaction which can be used to grow up small segments of DNA. With this technique Jeffreys had exploited mini satellite areas of DNA. This technique was used to specifically identify individuals from their DNA, a technique which has completely transformed forensic science and the identification of criminals.

Professor Goudie had used similar techniques to investigate immunoglobulin and T-receptor gene arrangements which are small round cells in the blood and can be divided into two groups, B-cells which are involved in the production of antibodies against infection, and T-cells which perform an entirely different type of immunological reaction and can kill some infectious reagents directly without the effect of an antibody. Both types of cell can be recognised by antibodies against immunoglobulin gene and T-receptor gene. This appeared very interesting and it was decided to pursue research along these lines.

Each case of lymphocytic leukaemia will have a different gene arrangement and this will be clonal. All malignant cells are clonal, that is they are derived from a single cell progenitor of the same genetic background. We decided to investigate minimal residual disease in childhood acute lymphoblastic leukaemia. It may be that a child is fit and well and has a normal blood and bone marrow but if looked at by sensitive DNA techniques evidence of residual leukaemia might be found.

It was therefore decided to pursue this line of investigation. DNA can be extracted from stored bone marrow slides and so the baseline immunoglobulin gene rearrangement may be determined. In follow up bone marrows the same technique would be used to look to find if there is any evidence whatsoever of disease in this bone marrow. The technique is extremely sensitive and very effective.

All the childhood ALL marrows at St James's were diagnosed in the Diagnosis Laboratory at Cookridge and were available for investigation. Caroline Shiach who had an interest in childhood leukaemia went to the follow up clinics at St James's Paediatric Oncology and brought follow up bone marrows for investigation.

The results proved very interesting and Paul Evans was able to go to the American Society of Haematology (ASH) to present his findings to the leading experts in childhood leukaemia in the world. This was a very significant achievement for Paul who came to us as a supermarket shelf stacker from Pontefract and had no university education whatsoever.

Within the Institute of Pathology it was the policy to look for areas of work that could be rationalised. It soon became apparent that there was great virtue in merging lymphoma histopathology from the Department of Pathology with the Cookridge Leukaemia Diagnosis Service and newly created DNA laboratory.

Steve Scott, the top grade Biochemist in charge of the Leukaemia Diagnosis Laboratory at Cookridge had written a book on the histochemistry of acute leukaemia, had obtained a PhD in this discipline and was awarded Fellowship of the Royal College of Pathogists based on his scientific publications. For a young man who came to us with O-levels this was a very distinguished career, but he realised that the future was in recombinant DNA technology which was not his field and when he was offered a very lucrative job in the United States he accepted it.

Steve Scott's departure was a great loss but in some ways it facilitated the merger of the three sections into a department which was called Haematological Malignancy Diagnosis Service (HMDS). The Head of the new service was Andrew Jack whose scientific training enabled him to successfully manage all disciplines. There was a problem however over funding the new Department because both the Lymphoma and Leukaemia Diagnosis Services were funded mainly by the Leukaemia Research Grant for Epidemiology and this was beginning to fade away. I therefore went to see Stewart Ingham, Chief Executive of the Trust, with my predicament and he arranged for bills to be sent out to the Chief Executives of hospitals using our services and all paid up immediately.

We therefore had a sound financial background for HMDS to continue into the future. I worked in HMDS several sessions a week reporting haematological bone marrows and films and then there were about 6,000 requests per annum. The figure is now over 26,000 requests per annum. The Department of Health now recognises HMDS as the blue print for any similar laboratory that should be set up in England.

The HMDS Department is staffed by many talented individuals; some started with University degrees, others did not. With support and encouragement they have produced work of a very high quality in the fields of flow-cytometry, cytogenetics, and molecular biology. They have produced many research papers of high quality, many in international journals of high repute. They have contributed very significantly to the reputation of haematology and pathology in Leeds.

I then looked at the overall state of Haematology and thought that compared to other departments in the Institute of Pathology that we needed a whole time academic presence. I therefore decided to apply for a Senior Lecturer in Molecular Haematology and had discussions with David Grant, Scientific Director for the Leukaemia Research Fund (LRF). This culminated in the production of a 60 page booklet printed and bound for me by the University Department of Histopathology as a form of application It gave background reports on medicine in Leeds and the School of Medicine, and the structure of the United Leeds Teaching Hospital Trust. There were descriptions of haematology in Leeds and the individual departments concerned.

When I began to prepare the application I assumed that the successful applicant would come along with his own projects and techniques and would plan his or her projects for the future. However the Leukaemia Research Fund insisted that I write in detail a proposed project for the successful applicant. This I did with considerable help from colleagues, and in particular Andrew Jack, and it was in essence an amplification of the work we were already doing in the DNA laboratory.

The project was basically an investigation of lymphocytosis (increase in number of lymphocytes) in patients by immunological and genetic techniques. There would be in the beginning, a screening of populations of patients, blood donors, and geriatric patients attending day clinics for a lymphocytosis. They would be subject to FACS analysis of lymphocyte populations and DNA analysis for clonality where indicated. Any abnormalities found would be then screened for autoimmune disease. Screening for abnormal populations of lymphocytes would then extended to patients attending clinics with autoimmune diseases. From this work it was hoped that the laboratory and clinical significance of clonality of T and B-cells would be determined but also throw further light on the evolution of lymphoid leukaemia and lymphoma.

The document was accepted by the Leukaemia Research Fund who then wrote to me to say that this project might not be acceptable to the successful candidate and would I give summaries of a further six projects that could be available.

We were then inspected by David Grant, and two senior research medical figures from the field of malignant haematology in England who appeared satisfied by what they saw in Leeds. A lunch was arranged with Professor Jewell, Dean of the School of Medicine, who when asked what help would be available from the Medical School for the successful candidate gave the answer "none".

I was however able to obtain a letter from Stuart Ingham, the Chief Executive of the Trust, confirming that he would arrange for the funds to employ the Senior Lecturer as a Consultant at the end of the grant.

At this stage I felt I was undergoing ordeal by fire and water similar to that suffered by Tamino in Mozart's opera the Magic Flute. In the end however I felt that the gods Isis and Osiris relented and David Grant in the person of Sarastro smiled upon us and awarded us over £600,000 to appoint Gareth Morgan, a Leukaemia Research Fund Fellow at the Royal Marsden Hospital. As I fully expected from the beginning Gareth came with his own ideas for research. All the projects I had suggested in the application were superfluous.

Over the past two or three years there had been more consultant appointments. In 1993 as the Institute was formed, Graeme Smith was appointed consultant as I was spending a considerable time on Institute affairs. Geraldine Bynoe had done her training as a part time trainee, was given two sessions: one to fill the post for a consultant session at Otley, and another one working in the Leukaemia Diagnosis Laboratory at Cookridge. Geraldine however got a full time consultant haematology post at Harrogate and retained one clinical session at the Infirmary. Graeme then took over the clinic at Otley.

Caroline Shiach meanwhile was appointed consultant and shared her duties between HMDS at the Infirmary and Paediatric Oncology at St James's. She did not stay long in this appointment however as she got a job as a haematologist in coagulation at Manchester Royal Infirmary and was able to join her partner in Manchester.

By this time Derek Norfolk applied for and was awarded the post of Director of Research for the Trust. Thus Haematology lost three more sessions. I was finding myself spending more and more time reporting slides in HMDS. I went to see Stuart Ingham, the Chief Executive, and asked for three sessions to compensate for Derek's and Caroline's loss. He at first refused outright but I put pressure on him because this would also limit the time available to me for my responsibilities as Director of Pathology. I therefore had five sessions which enabled me to merge them with six sessions that Dr

Mike Galvin had available at Pinderfields Hospital, Wakefield, for a consultant.

These sessions were very important to me because Pete Hillmen, one of our senior registrars, had just passed his final MRCPath in Haematology and was available to apply for consultant posts. As I have already said in a previous chapter, Pete when working at the Hammersmith as a Research Fellow had found the cause of the rare condition paroxysmal nocturnal haemoglobinuria (PNH) and had produced several good research papers since. He was able to apply therefore for this post between the two hospitals and was successful much to my delight. Due to the efforts of Pete a National Centre for PNH has been set up in Leeds.

Apart from winning virtually every golf tournament he enters, Pete has continued to do very successful research particularly in the field of chronic lymphocytic leukaemia, and recently the University have appointed him as a full time Professor of Experimental Haematology in the University.

Meanwhile things were progressing in the Pathology Institute. There were problems in pathology at Bradford where the Trust were quite prepared to privatise the pathology services. As an alternative it was suggested that there should be discussions between Leeds and Bradford about merging the two pathology services. As the Consultants at Bradford had got extensive private practises they showed some reluctance, but eventually the two Trust Chairmen got together and a decision was made that the merger should in fact happen. One of the features of this merger was the fact that a large food laboratory at Morley near Leeds, owned by ASDA, and superbly fitted out, was deemed surplus to their requirements and it was therefore leased on behalf of the two Pathology Departments. It was used basically by both departments for non acute services such as exfoliative cytology which were rationalised between the two Centres.

Thus the Institute went from strength to strength. It never overspent and indeed the same can be said for the Institute of Dentistry.

The Trust therefore was very impressed with this system of management and decided to extend it to the rest of the hospitals. This was met by some reticence in the first place but nevertheless it went ahead with the Units being based on the previous Divisional Structure, for example, a Division of Surgery and a Division of Medicine.

At this time moves were made to merge St James's Trust and the United Leeds Teaching Hospital NHS Trust at the Infirmary. There was a vote among medical staff for this and the merger was approved. Stuart Ingham had ideas therefore that the system of management by divisional groups could be extended in all of Leeds, for example, a common Institute of Pathology. The trouble was however that John Major lost the next election and that was the end of that. Stuart Ingham, the Chief Executive at the Infirmary, was "Thatcherite" in his views and very much disliked by the local Labour MP's. Although Stuart was an innovative executive who could have managed the new entire Leeds Trust he was however sacked, presumably because of his political views. The Institute was therefore left to the mercy of the new administration. I retired just in time.

CHAPTER 12

"THE FRIENDS ...."

"and now abideth faith, hope, charity, these three, but the greatest of these is charity"
Bible, Corinthians 13.13

The clinical work on the Professorial Medical Unit began to grow and treatment became more successful as newer treatment, antibiotics, and an improved supply of blood products became available. It soon became apparent to patients and their relatives, as well as the doctors, that the treatment of leukaemia on an open ward was unsatisfactory and informal meetings of visiting relatives took place which ultimately led to the formation of the group known as The Friends of the Leukaemia Unit.

After a series of preliminary meetings The Friends of the Leukaemia Unit was set up as a Charitable Trust and the first meeting of the Trustees was held on 28 September 1979.

Members of the Trustees
Friends of the Leukaemia Unit

M H Maufe	1979 - 1997
B G Stenhouse	1986 - 2004
K W Stenhouse	1979 - 1985
G M Smith	1993 -
R G B Bucknell	1979 - 1983
A John	1995 - 1997
J A Child	1979 - 2005
A Thompson	1997 - 2012
B E Roberts	1979 -
Mathew Baker	2000 -

B A Atkinson	1983 - 2001
R Owen	2005 -
J Barritt	1983 - 2008
Mark Baker	2010 - 2013
T Davis	1985 - 1986
D Fox	2013 -
D R Norfolk	1986 -

At the first meeting rules were laid down as were the objectives. The main objectives were:

1   To set up a ward for the treatment of patients with leukaemia.

2   To provide furniture and equipment for such a ward.

3   The promotion of research into leukaemia.

4   Advice and assistance to patients with leukaemia.

The second meeting was held on 18 January 1980. It is interesting to note that already some £6000 had been collected and this was held on deposit at an interest rate of 15%.

There were at this time other individuals who wished to become Trustees. It was decided to appoint Trust Managers and the three appointed were Mr A Addison, Mr G S Share, and Mr J B Bone. Mrs Sue Natkus and Mr Johnson would soon be added to this list.

The development of bone marrow transplantation was now under active consideration in the country but this would have to be a Regional project. At a meeting in September 1980 there was a proposal to alter the side room on Ward 29 in the Martin Wing as the provision of an entirely new ward appeared to be several years away. It appeared that the cost of modifying the four bedded side room on Ward 29 would be between £30,000 and £40,000. It was hoped that the hospital would bear some of the cost.

In July 1981 it was proposed that the side ward be converted into a three bedded ward with a separate cubicle with en-suite facilities. These plans however were put on hold until Leeds plans for bone marrow transplantation were announced. In October 1981 it was announced that the Yorkshire Region would provide funding for a transplant programme in Leeds. This would mean support for the four bedded unit and extra nursing staff. At a meeting in April 1982 it was announced that the building would go ahead and that The Friends of Leukaemia Unit would contribute £35,000 for rebuilding and equipping the side room in Ward 29, a total cost of £90,000. There were still no facilities for Day patients; they had to borrow beds on the ward.

The opening took place on Saturday, 23 October 1982 at the Infirmary. The opening was marred by the death of Sir William Tweddle, Chairman of the Regional Health Authority, who collapsed and died at 9.00 am on the very morning of the opening ceremony. Mr Robin Wood, Chairman of the Leeds Western District Health Authority deputised. The new facilities proved to be a considerable help in the organisation of the management of leukaemia.

At the next meeting of the Trustees the resignation of Brian Bucknall, due to work commitments, was announced. After discussion two new Trustees were proposed and accepted: John Barritt, Finance Director of Schofields Department Store in Leeds, and Mr Brian A Atkinson, a District Inspector of Taxes. It was further noted that the total income of The Friends was now £100,000.

On 13 August 1985 Ken Stenhouse, a retired Bank Manager, died much to the dismay of his fellow Trustees. At the meeting of the Trustees in 1985 Trevor Davies was elected to the Board of Trustees and agreed to accept the post of Secretary and Treasurer. Unfortunately Mr Davies was moved by his employers to Newcastle in 1986 and had to resign at short notice. He was replaced by Mrs Betty Stenhouse, widow of Kenneth Stenhouse, who in fact took over his role as Secretary/Treasurer. Betty was to play a major role in the function of the Trust for years to come.

Meanwhile the work of the Clinical Unit continued to expand. Sixteen to twenty transplants were being performed at St James's and the accommodation at the Infirmary was now very limited. The

Friends were informed that substantial funds would be required in the future as the search for a new accommodation had begun.

In 1988 a major development took place; the creation of a post of Chemotherapy Nurse. This nurse would be of enormous help to nursing staff and doctors. It was agreed that The Friends would fund the post for three years with the hope that the hospital would fund it after this. Sister Bilborough, the Sister in Medical Out Patients, agreed to take the post; she was performing many of the tasks required of the new appointee at the present time. Sister Bilborough, now retired, is the new Secretary to the Trustees.

In late 1989 it was proposed by the hospital that the new haematology beds should be situated on D Floor, Brotherton Wing. It was noted at this time that The Friends had approximately £240,000 in various accounts and much of it was invested at an interest rate of between 13% and 14%.

It was however decided to hold a major appeal to fund this development. A sum of £250,000 was initially sought and a new account for this appeal fund was opened. A brochure for the appeal was prepared. The cost was now estimated to be between £335,000 and £350,000.

In general, the appeal to business in Leeds and various charitable trusts was disappointing, but various donations on an individual basis boosted the funding. In particular, Richard Grantham's, a retired policeman, 999 Appeal raised £45,000 and there were further donations from Joe Longthorn, the singer/impersonator and patient of Dr Child, from a charity concert, and the Store singer in Leeds, Danny Freeman, also made a significant contribution.

The Members of the Management Committee also raised a very significant amount of money. Mrs D Bramble from the East Coast, a farmer's wife, Mrs Sue Natkus representing the Tadcaster region, Mr Geoffrey Share from Pontefract, and Mr C Johnson from Scarborough, all were very active in their localities.

204

The Friends were also very helpful to the Laboratory Diagnostic Services in Leukaemia. When the Leukaemia Diagnostic Service was first set up at Cookridge, The Friends were very helpful in the purchase of equipment including an Apple computer. The provision of this Leukaemia Diagnostic Service was very helpful in the acquisition of a major research grant awarded to study the epidemiology of leukaemia and lymphoma. Although The Leukaemia Research Fund paid for all reagents, The Friends purchased relevant equipment, also funds for research projects which were not associated with The Leukaemia Research Fund. It also became apparent that a FACS machine (fluorocein automated cell sorter) was required. The machine cost almost £100,000 and The Friends were unable to afford this, but did agree to fund the running costs if the Regional Health Authority would fund the machine.

The volume of work had now increased so that space was very cramped in the laboratories at Cookridge, but local portocabins became available and The Friends purchased these. The portocabin laboratories were taken over in 1988 with a plaque on the door indicating that these laboratories belonged to The Friends of the Leukaemia Unit and it was decided to hold an open day, which was very well attended, for members of The Friends and contributors to come and see the new laboratories and all the equipment purchased by The Friends.

In February 1992 HMDS was created and this incorporated the new DNA laboratory. The DNA work had expanded and The Friends of the Leukaemia Lymphoma Unit contributed £50,000 to laboratory equipment. The Friends also paid for the annual salary of the Senior Registrar, Dr Caroline Shiach, for a period of one year. The overall funding of HMDS contributed largely to the appointment of Dr Gareth Morgan as a Leukaemia Research funded Senior Lecturer in Molecular Haematology at the Unit. The investment by The Friends in HMDS paid rich dividends.

By June 1992 the total sum of money in the Appeal Fund was £315,000. It was at this point that it was agreed to rename the title of

The Friends to The Friends of the Leukaemia and Lymphoma Unit (LGI).

The ward was finally opened by the Earl of Harewood on 20 March 1992 at 2.00 pm. The Earl of Harewood had previously agreed to become a patron of The Friends. It was also agreed that a reception would be held in the Civic Hall for members, donors, and fund raisers.

Thus the Department of Haematology had a modern state of the art ward with 14 cubicles including two isolation rooms. This would aid the treatment of leukaemia and lymphoma; permit autologous bone marrow transplantations, and promote the introduction of peripheral blood stem cell transplantation in leukaemia, lymphoma, and myeloma.

Meanwhile clinical activity continued to increase. The main feature of clinical research, apart from therapeutic trials, was the development of peripheral blood cell transplants. The blood was processed in the Blood Transfusion Department and The Friends funded a technician for the Blood Transfusion Department to assist with the project. The Friends also funded a clinical research fellow/transplant co-ordinator for this project.

A major development in patient care took place with the availability of the Orthopaedic Out Patient Department which was situated opposite the ward on D Floor, Brotherton Wing. It would provide facilities for the out patient management of patients. There would be four beds for out patients receiving chemotherapy and blood transfusion, and a room for stem cell harvesting. The Out Patient Suite was finally completed in August 1999 at a cost of £157,000, funded by The Friends. The opening was carried out by the Calendar Girls who also made a donation of £50,000 to The Friends. Mr Matthew Baker, is a solicitor, and his father, John Baker, was a patient on the Unit and was the subject of the film Calendar Girls. His mother played the role of Miss February, and his mother-in-law the role of Miss July. Matthew was approached to become a Trustee and take over the Treasurer's post, with Mrs Stenhouse continuing as Secretary.

At this stage HMDS made a further request for money to perform the FISH (fluorocein in situ hybridization) technique to determine chromosome abnormalities in leukaemia and lymphoma and received a further £20,000.

It was now apparent that the new Oncology Wing at St James's was going ahead and all haematology patients, plus HMDS, would be transferred there. This did not stop The Friends continuing to support more research projects and fund clinical research.

The Department of Haematology is now established as a department with an international as well as national reputation. This has been built up through the years and it is clear from this review that this would not have been possible without the enormous financial contribution of The Friends. Bexley Wing at St James's provides state of the art facilities for the management of patients but if scientific and therapeutic developments are to continue then The Friends will be needed as much in the future as they have been in the past.

The opening of the new Leukaemia Unit on Ward 29, the General Infirmary at Leeds. From left to right: Mr Michael Maufe, Chairman, Friends of the Leukaemia Unit; Sister Freda Ellis; Mr Robin Wood, Chairman, Leeds Western District Health Authority; Dr Tony Child, the author.

The opening of the portacabin laboratories at Cookridge Hospital with presentation of cheques to the Friends.

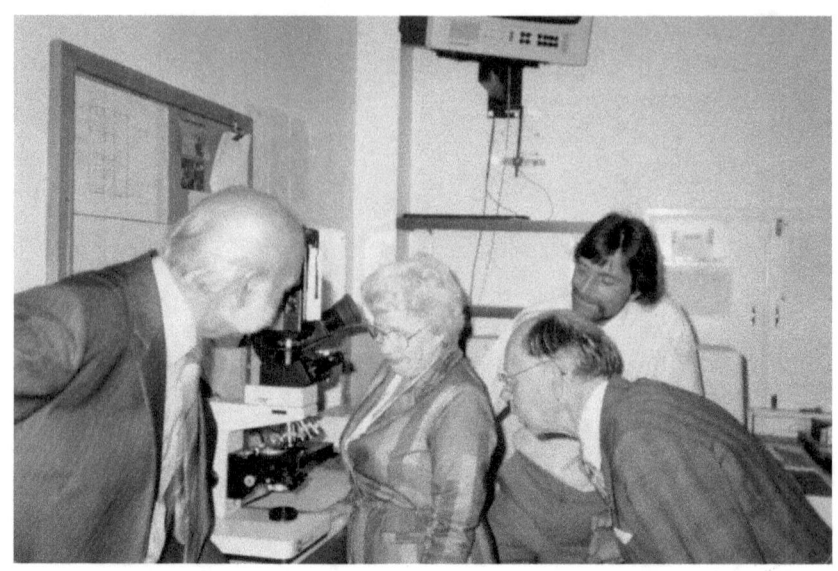

Mr Geoffrey Share, Mrs Betty Stenhouse, and Mr Michael Maufe look at a microscope provided by the Friends. Dr Steve Scott, Top Grade Scientist looks on.

The 7th Earl of Harewood, Patron of the Friends.

Mrs Betty Stenhouse, Secretary of the Friends of the Leukaemia Unit, at her retirement presentation in the Civic Hall, Leeds.

Mrs Carol Bilborough has served the Leukaemia Services for over 40 years, first as a Nursing Sister and now as Secretary to the Friends.

Party time at the Infirmary (1958). Mr Myles Gibson, Senior Registrar in Neurosurgery, is on the bicycle, the author is standing behind him. Miss Pat Jessop, who would have a distinguished career in Nursing is on the left. An Austin 7 has been manhandled through the front door of the Infirmary.

214

INTRODUCTION

Welcome to the first addition of Path News, a newsletter for all members of the ULTH Pathology Institute. There has been much criticism of failures of communication within the Institute. Perhaps this is a chance to reverse the trend!

One of the most important features of this newsletter is its editorial independence. We see ourselves as more akin to *Private Eye* than *Pravda* and you will note that only one of the Editorial Committee is a member of the Pathology Institute Management Board. We are committed to making this newsletter an open forum for news, views and debate but this can only be achieved with your support and participation. The newsletter will be published quarterly in the first instance and we invite contributions and correspondence from any member of the Institute. Those of you with an artistic or journalistic bent would be welcome to join the Editorial Team.

The allusion in the title of this newsletter will be immediately obvious to older members of the Institute (if you were born after 1965 ask an older and wiser friend). All successful journals need a memorable title therefore, true to the principals of the market economy, we announce a competition to provide us with a suitable name for the newsletter which will be used in subsequent issues. There is a prize for the best entry received (for details see page 4).

We hope this newsletter will become a valuable part of the life of the Institute. By bringing news and updates together in one journal we may even save the occasional rain forest. The success of this venture rests entirely with our readers. We hope you will use this opportunity to the full.

WELCOME!

MESSAGE FROM THE DIRECTOR
Bryon Roberts

If the Institute of Pathology is to progress it must have shared objectives and shared values. To achieve this, good communication is paramount but is often difficult to accomplish. I am, therefore, very grateful for the efforts of the Editorial and Production Team of *Path News* in helping to fulfil one of the major aims of the Institute.

The first issue of a newspaper produced by the Institute of Pathology. A caricature of the Director (the author) is on the front page.

CHAPTER 13

"PARTY TIME"

All hard pieces of work should be followed by a party and this book is no exception.

All young doctors upon qualification have to spend at least one year in hospital working under supervision before they can become registered. This meant that young doctors aged around 24, mainly men, were taken into a hospital which was full of nubile young women, there was free booze and, in effect, the key was thrown away. They were all warned that there were strict rules to be obeyed and that should any young doctor be found in bed with a young nurse then both would be instantly dismissed. However it is worth noting that no such offence had been recorded within living memory. It was inevitable with continuous hard work under pressure that parties would ensue and The General Infirmary at Leeds was no exception.

The first major party was traditional and celebrated the in-coming House. The main feature of this was a trolley race between house physicians and house surgeons which took place along the long main corridor of the Infirmary under the influence of drink. There was a similar party after six months at the change over of the house and this again followed the same pattern. I remember very well Sister Busby, a ferocious night sister, coming onto the main corridor and berating party members only to be picked up by her arms and legs and carried from one end of the corridor to the other.

At the end of the corridor leading to Thorseby Place and the Medical School there is a large archway where, in days of old, carriages came into the neighbouring quadrangle to allow patients to be taken into the operating theatres which were then situated in the Littlewood Hall. It was above that archway that Matron's apartment was situated and the archway beneath had two large doors converting it into the garage for Matron's car. Somehow someone managed to open the doors and Matron's Morris Minor was manhandled up a flight of steps and taken to the middle of the corridor. There were no repercussions from this, as far as I know, and Matron like many others turned a blind eye to what went on at the resident's parties.

In fact Mr Tunstall, the House Governor and Secretary to the Board or, in modern terminology, the Chief Executive, would come in on a morning and say to Bill when he saw the remains of a party scattered around "did anything go on last night, Bill", to which Bill would reply "I saw nothing, sir".

There were other parties to celebrate then such as birthdays, stag nights, and any excuse that one could think of.

The residents mess was, every night, lit with subdued lighting. There was a bar and music played. The music was usually that of Harry Belafonte and every time I hear the words of the calypso, which goes as follows:-

"Down the way where the nights are gay,
and the sun comes shining 'oer the mountain tops,
I took a trip on a sailing ship
and when I reached Jamaica I made a stop."

It takes me back with fond memories to my life as a resident in The General Infirmary at Leeds.

Around Christmas in hospital there are several major events. This day, for me, consisted of a ward round followed by a visit from the Consultant and then I carved the turkey for the patients who remained in hospital, most of whom were unfit to eat. Then it was considered the duty of every houseman to go and act as waiter and serve lunch for the resident domestic staff.

There were three major events in hospital at Christmas which were in the form of pantomimes. There was a resident's pantomime held in the Nurses Home, the sister's pantomime held in the Nurses Home, and the medical student's pantomime held in the Students

217

Union. For the resident's pantomime I had to dress as a woman patient and wear a large ginger wig. Several minutes before I was due to appear on stage I got a call to say that I had a patient admitted with bleeding oesophageal varices. I disposed of my wig, disappeared to see the patient with a white coat round my dress, inserted a needle, got a saline drip going, ordered the blood, and departed back to take my place in a bed on stage just in time.

The sister's held a pantomime the title of which was "Bleep, Bleep, housey man". The basic theme was based on the fact that the junior medical staff had acquired electronic bleeps at the beginning of their new jobs in September. The songs were all sung for some reason in a Chinese accent and I am afraid that I did not escape being incorporated into one of the songs and the verse went as follows:-
"Bryon Roberts very sleepy, he likes his bed,
When old Bill comes up to wake him, he like lump of lead."

I cannot argue about this because it was absolutely true.

The medical student's pantomime that year caused an outrage. The fact that Hugh Garland, the Consultant Neurologist, had installed his mistress by the nickname "Tweetie" into the Infirmary as Resident Medical Officer, was well known throughout the medical world in Leeds. There was at that time a popular song sung by Mel Torme in which a canary in a cage, called Tweetie Pie, is threatened by a huge cat known as Sylvester. The words of the students modified this song to begin:-
RMO: "I am a little RMO, my name is Tweetie Pie.
Consultant: I am a great big consultant, my name is Hughie G".

Dr Garland was present at the concert, as many consultants were, and were prepared to accept any insult, but this was too much for Garland. He considered this an insult to his mistress Tweetie and he refused to teach for several months.

In a previous paragraph I have described various systems of management, including line management and team management, but

Arnold Tunstall the House Governor, believed in management by the products of the grape and the grain. When consultants asked to see him about some problem they were encountering he would assess the degree of severity of the problem. At the beginning of the interview Arnold would complain that his artificial leg was rusty and in need of lubrication and he would proceed to the drinks cabinet where, if the problem was mild he would pour a sherry, or if the problem was severe and the consultant was red in the face with steam coming out of his ears, he would receive a stiff gin and tonic. After a few minutes most problems would seem to pass away and the consultants would leave his office in a bemused state.

This principle applied to some other meetings, in particular the meeting between the Board of Governors and Officers, with the top hierarchy of the University to discuss the knock-for-knock financial agreements. This meant balancing between the cost of providing facilities for teaching students against costs of the University providing services to the Infirmary. There were insufficient statistics to put this onto a logical basis and arrangements were always informal. It would so happen that the morning session was acrimonious and bad tempered but this would be then followed by lunch. Lunches were very high quality and were provided by the hospital kitchen and I remember very well attending lunches with large legs of cold ham and intact poached salmon, with many other delicacies. Lunch would begin with sherry or a gin and tonic and after high quality white wine a very good quality claret was supplied. The Infirmary with its endowment fund kept a very good cellar. Lunch would be completed with cheese and a vintage port. The meeting then would commence in the afternoon and it was said that a proposal would be made by one side or another and in a very slurred voice that things should remain the same as last year. There would be unanimous agreement and then the members of the meeting would all be taken home.

Another major function of the Board of Governors was to judge the winner of a competition for the best decorated ward in the hospital. Before this competition, when I was the House Physician on Ward 7, I was instructed by Sister to get a certain patient in for further investigation. This patient had atrial septal defect and remained quite well but on admission he was placed in the middle of the ward and supplied with paint brushes and large sheets of paper on an easel and he was expected to provide decorations for the ward. This he appeared to do on an annual basis. After much hard work and a few non invasive investigations he was allowed home.

The week before Christmas the Board of Governors would do a large extensive ward round, get a drink at most wards, and make a judgement as to the winner of the best decorated ward competition. Once again they were all taken home.

Early in November 1973 the City of Leeds decided to award the Freedom of the City to the aircraft carrier Ark Royal. During the war the Ark Royal had been sunk and the people of the City of Leeds collected enough money to build another one. There was therefore a close bond between the City of Leeds and the Ark Royal. The Ark Royal was in the South Sea islands on its tour of duty and held what they called a garden party and made a collection of money which they were going to donate to the Infirmary.

When the hospital received news of this gift it decided to place money out to bids and a person or a department could apply to receive this money. I discussed this with Ken Major in Blood Transfusion who was keen to acquire a machine for washing frozen red cells. At the time frozen red cells were in vogue for transfusing patients who had white cell antibodies. The process of freezing and thawing kills all white cells. The red cells however are frozen in glycerol as a preservative and have to be washed thoroughly before administration to a patient. We therefore made a written application for the funds to supply this piece of equipment and much to our surprise were successful.

As recognition of this gift Arnold Tunstall organised a party for naval ratings in the hospital to which Ken Major and I were invited. The drink flowed in abundance that night until at a certain stage the Infirmary stores of booze ran out. Arnold then woke up the Hospital Mess secretary to open up the residents bar for the party and so the celebrations continued until the early hours. Fortunately my wife came down with my young son to pick me up and my son can remember that party to this day. The next day from the vantage point of the Brotherton Wing we watched the naval ratings, ashen faced and with fixed bayonets march past the Queen Mother who was taking the salute ,standing in a balcony on the Civic Hall.

Departmental parties were quite common in the Infirmary but there was very little cause for celebration in the Department of Pathology where Professor Lumsden, a dour Aberdonian Scot, did not believe in Christmas and celebrated the New Year. However we managed to get Department of Haematology parties going in the early 1960's and held a collection to provide drink from everyone on a weekly basis throughout the year.

The General Infirmary at Leeds has a neo-gothic style of architecture and one limb of the Department of Haematology had suitably shaped windows to provide a gothic theme for our first party. The windows had been covered with black sheets of paper with slits to let the light in and to simulate the bars of a prison. Strings of cotton dangled from the ceiling, as did various cut out bats. Finally Charles Buchan, the Senior Chief, had a word with his colleagues among the electricians and they came and fixed up some subdued UV lighting. So our main theme of converting the Department of Haematology into a cave of bats went very well. The party was enjoyed by all. At this time the Infirmary and St James's worked alternate nights on acute so there was one very busy night and then a night of relative peace.

On one of these quiet nights before Christmas all the departments would hold their Christmas parties. The entire building would be lit up with the sound of music, and people would wander from department to department to take a drink. There was an understanding that the various Work Departments were invited to all these parties to thank them for the work they did during the year. In those days there were no forms to fill in and no paper work. All the maintenance work was done by word of mouth between the Senior Chief and the foreman of that particular section of the Works Department.

For the next party it was decided to hold a Greek symposium and for this we needed the statue of a Greek nymph. Cedric Abbot, the then Registrar (who would later become a consultant) had a bright idea. He approached Beryl Walsh, a young woman in the photography department and asked her to be a model. He went to the A&E Department, collected stockinet and plaster-of-paris and made a mould of Beryl (the intimate details of this have never been revealed) and we ended up eventually with a statuette of a Greek nymph which adorned the department to everybody's great satisfaction. After the party this Greek goddess was taken and mounted above the front door of the Infirmary and later sent to the Works Department of the Infirmary where it survived for several years.

In due course however it was not possible to hold parties within the department as we began to acquire bits of expensive equipment which could be easily damaged. We therefore hired the Garland Gallery next to the Littlewood Hall in the Instructional Block to have our parties. The major theme was to have a beer brewing competition among groups of the staff and I would be the judge. They would buy the kits from Boots the Chemist and brew the beer in large plastic containers which had previously contained Isoton, a balanced saline solution for use in Coulter counters

One morning around party time I entered the Infirmary through the consultant staff entrance at the front door, handed my keys to the parking attendant, picked up my newspaper, and strolled with an air of dignity through the entrance hall covered in paintings and sculptures of distinguished physicians and surgeons from the past. I was struck however by the fact that the whole place smelt like Tetley's Brewery. This was difficult to understand but as I walked up the corridor and got to the stairs leading up to the bust of Lord Moynihan I could see rivulets of beer descending from the other side of the basement corridor. It then dawned on me what had happened. The Isoton containers made of thin plastic in which the beer was brewed had burst and liberated the beer into the cellars underneath the Department of Haematology and down to the main corridor of the Infirmary.

I waited for a telephone call from Richard Oswald, the District Administrator, and thought I was "for it" in a big way. However I never heard a thing. The whole hospital stank of beer but no comment was made by anybody at any time.

The technical staff were quite fond of participating in inter-hospital pub quizzes. Very often fancy dress would be employed and on one occasion a technician by the name of Robert decided to go dressed as a Viking with a helmet with two horns and a sheepskin around his shoulders. As the night wore on Robert became much the worse for drink but refused to surrender his keys to Charles Buchan his boss. He drove home, entered a field through a dry stone wall and arrived home too scared to enter his house and see his wife, something of a harridan. He decided therefore to sleep in the caravan in the drive. Next morning his wife awoke and became aware that there was no husband beside her. She jumped out of bed, drew the curtains, and looked out of the window just in time to see a Viking leaving the caravan and asmashed up car in the drive. The immediate consequences are not known but it is said that certain privileges were withdrawn for quite some time!

Eventually within the hospital restrictions were put upon parties held within departments because of the noise and disruption. We therefore arranged to have parties outside and one of these I remember very well. The organisation was put in charge of the "Shad" whom I have described before. He organised for a party in a restaurant in the basement of Joseph's Well, an office block adjacent to the Infirmary. He had also arranged for some entertainment, so as we sat down an oriental Fakir came on stage and proceeded to walk on burning coals. This was quite impressive.

However for the next turn I was called onto the stage and realised this was the time for the ritual humiliation of the boss. I was asked to stand against the wall, arms outstretched, in front of a large board covered in paper and then a large Red Indian appeared on the stage with his hands full of machetes. He proceeded to hurl them towards me and cut out my silhouette. I was very pleased to survive this episode.

223

Parties were also held in our peripheral laboratories, such as the laboratory at Cookridge. At one of the Cookridge parties the Shad, who was known to have access to soft drugs, made a cake which contained the drug Magic Mushrooms. No one was aware of this, the cake was eaten, and it was said that the party that followed was the best ever. And so the parties continued but as time went on more and more of these began to be in the form of a dinner dance at an hotel. There was also the additional threat of the breathalyser.

When I came to retire in 1998 there were three parties that I remember so well. The first departmental party was held in the Banqueting Suite at Elland Road the home of Leeds United Football Club. This again was a very memorable and official event. HMDS had organised in the previous week a separate party at Roundhay Park Mansion. I remember when I was asked to attend this party I asked what the party was for and what was it all about. I was told it was to celebrate my retirement and for the first time I realised that my departure was imminent.

At the end of my last working day I was reporting slides in HMDS when half way through reporting a patient with a mild lymphocytosis, Steve Richards appeared and said "we want you in the next room". He would not allow me to finish the report but took me next door to a further final celebratory party where they presented me with a book full of photographs of previous parties which I treasure very much. This however was the last time that I would look down a microscope in anger, as it were, after the the completion of 47 years of study and service in the General Infirmary at Leeds.

CHAPTER 14

POSTSCRIPT

In Chapters 10 and 11 the major upheavals in the Medical School were described with its knock-on effects on the Pathology Service. Was all this worthwhile?

The next Research Assessment Exercise was in 2001 and the results exceeded the expectations of the School of Medicine.

The assessment panels reviewed Unit Submissions and awarded ratings on a scale of 1-5*. A rating of 5 equates to attainable levels of international excellence in up to half of the research activity submitted and to attainable levels of national excellence in virtually all the remainder. Unit of Assessment 1 (Clinical Laboratory Sciences) including Molecular Medicine, Pathology, and Medical Physics scored 5. All received high praise for the high proportion of research being of international quality.

Thus the Institute of Pathology, a National Health Service Unit run by doctors, scientists, and technicians, with support from an outstanding Finance Manager and Personnel Manager, managed to:-
- never exceed its budget but managed to grow and increase its staff;
- always met its cost improvement target;
- have the lowest absentee rate in the Trust;
- have the largest research income in the Trust;
- and achieve a score of 5 in the Research Assessment Exercise.

Thus Pathology was where it was when I was at Medical School: At the top of the tree!

www.ingramcontent.com/pod-product-compliance
Lightning Source LLC
Chambersburg PA
CBHW060832170526
45158CB00001B/143